Reducing the Seismic Vulnerability of Existing Buildings Assessment and Retrofit

Reducing the Seismic Vulnerability of Existing Buildings Assessment and Retrofit

Special Issue Editors

Tiago Miguel Ferreira
Nuno Mendes
Rui Silva

MDPI • Basel • Beijing • Wuhan • Barcelona • Belgrade

MDPI

Special Issue Editors

Tiago Miguel Ferreira
University of Minho
Portugal

Nuno Mendes
University of Minho
Portugal

Rui Silva
University of Minho
Portugal

Editorial Office

MDPI

St. Alban-Anlage 66

4052 Basel, Switzerland

This is a reprint of articles from the Special Issue published online in the open access journal *Buildings* (ISSN 2075-5309) from 2018 to 2019 (available at: https://www.mdpi.com/journal/buildings/special_issues/Seismic_Vulnerability)

For citation purposes, cite each article independently as indicated on the article page online and as indicated below:

LastName, A.A.; LastName, B.B.; LastName, C.C. Article Title. *Journal Name* **Year**, *Article Number*, Page Range.

ISBN 978-3-03921-257-6 (Pbk)
ISBN 978-3-03921-258-3 (PDF)

Cover image courtesy of Tiago Miguel Ferreira.

Contents

About the Special Issue Editors

Tiago Miguel Ferreira received his PhD in Civil Engineering from University of Aveiro. His research domains include the seismic vulnerability assessment of old buildings, the evaluation and mitigation of urban risks, and the structural safety assessment and retrofitting of ancient stone masonry buildings. He is currently a Researcher at the Institute for Sustainability and Innovation in Structural Engineering (ISISE) of the University of Minho, invited Assistant Professor at the Department of Civil Engineering of the University of Coimbra, and Associate Editor of *Conservar Património*, a journal indexed by SCOPUS and Web-of-Science and dedicated to heritage studies.

Nuno Mendes received his PhD in in Civil Engineering from University of Minho (Portugal). His research interests include the inspection, diagnosis, non-destructive testing, dynamic tests, advanced numerical modeling (FEM and DEM), and strengthening and monitoring of masonry buildings. Currently, he is a Researcher at the Institute for Sustainability and Innovation in Structural Engineering (ISISE) of the University of Minho. He participated in several consultancy projects and research works on the seismic vulnerability of historic buildings, such as Qutb Minar (India), Canterbury Cathedral (United Kingdom), San Sebastian Church (Philippines), Municipal Theatre of Rio de Janeiro (Brazil), Loka-Hteik-Pan Temple (Myanmar), the Presidential Palace (Portugal), and Carmo Convent (Portugal).

Rui Silva received PhD in Civil Engineering from University of Minho and KU Leuven. His research interests include the characterization, modeling, strengthening, and repair of earth constructions, as well as the development of sustainable building materials based on geopolymeric binders, and construction and demolition waste. He is also Principal Investigator of the project SafEarth, dedicated to the seismic protection of earthen built heritage. Currently, he is a Researcher at the Institute for Sustainability and Innovation in Structural Engineering (ISISE) of the University of Minho and member of the committee RILEM 275-TCE: Testing and characterization of earth-based building materials and elements.

Preface to "Reducing the Seismic Vulnerability of Existing Buildings: Assessment and Retrofit"

The devastating seismic events occurring all over the world continue to alert the scientific, technical, and political communities of the need to identify assets at risk and develop more effective and cost-efficient seismic risk mitigation strategies. Significant advances in earthquake engineering research have been achieved with the rise of new technologies and techniques, with potential uses in risk assessment, management, and mitigation. Nevertheless, there is still much to be done, particularly with regard to existing buildings, most of which have been built without anti-seismic provisions. The wide variety of construction and structural systems, associated with the complex behavior of their materials (raw earth, timber, masonry, steel, and reinforced concrete), greatly limit the application of current codes and building standards to the existing building stock. To tackle this global issue, there is a fundamental need for developing multidisciplinary research that can lead to the development of more sophisticated and reliable methods of analysis, as well as to improved seismic retrofitting techniques compliant with building conservation principles.

This book intends to contribute to the aforementioned goal by stimulating the exchange of ideas and knowledge on the assessment and reduction of the seismic vulnerability of existing buildings. Included are 10 high-quality contributions authored by international experts from Italy, Portugal, Morocco, Nepal, Czech Republic, and Spain. All contributions focus on the protection of existing buildings by considering the most updated methods and advanced solutions emerging from different fields of expertise.

Tiago Miguel Ferreira, Nuno Mendes, Rui Silva
Special Issue Editors

buildings

MDPI

Editorial

Reducing the Seismic Vulnerability of Existing Buildings: Assessment and Retrofit

Tiago Miguel Ferreira *, Nuno Mendes and Rui Silva

ISISE, Institute of Science and Innovation for Bio-Sustainability (IB-S), Department of Civil Engineering, University of Minho, 4800-058 Guimarães, Portugal; nunomendes@civil.uminho.pt (N.M.); ruisilva@civil.uminho.pt (R.S.)
* Correspondence: tmferreira@civil.uminho.pt; Tel.: +351-253-510-200

Received: 11 June 2019; Accepted: 17 June 2019; Published: 19 June 2019

Devastating seismic events occurring all over the world keep raising the awareness of the scientific, technical and political communities to the need of identifying assets at risk and developing more effective and cost-efficient seismic risk mitigation strategies. Significant advances in earthquake engineering research have been achieved with the rise of new technologies and techniques with potential use in risk assessment, management and mitigation. Nevertheless, there is still much to be done, particularly with regard to existing buildings, most of them built without anti-seismic provisions. The wide variety of construction and structural systems, associated with the complex behaviour of their materials (raw earth, timber, masonry, steel and reinforced concrete), greatly limit the application of current codes and building standards to the existing building stock. To tackle this global issue, there is a fundamental need for developing multidisciplinary research that can lead to the development of more sophisticated and reliable methods of analysis, as well as to improved seismic retrofitting techniques compliant with buildings conservation principles.

The present Special Issue of Buildings intends to contribute to the aforementioned goal by stimulating the exchange of ideas and knowledge on the assessment and reduction of the seismic vulnerability of existing buildings. As outcome, 10 high quality contributions authored by international experts from Italy, Portugal, Morocco, Nepal, Czech Republic and Spain were published. All contributions pursue the protection of existing buildings by considering the most updated methods and advanced solutions emerging from different fields of expertise.

Ferreira et al. [1] starts by presenting a comprehensive review of the most relevant vulnerability assessment methods applicable at different scales, as well as the most significant traditional and innovative seismic retrofitting solutions for existing masonry buildings. The authors highlight the need for a proper balance between simplicity and accuracy when selecting the most appropriate seismic vulnerability method or technique, and stress the importance of adopting vulnerability indicators that can be easily understood and interpreted, not only by the technical and scientific community, but also by citizens and governmental and civil protection authorities.

Mascort-Albea et al. [2] proposes the development of a set of documents for the evaluation and diagnosis of the state of existing buildings and infrastructure regarding seismic activity in Andalusia. The authors establish two specific protocols. The first, a short-term guideline, which enables the classification of damage and risk levels, and the determination of what immediate interventions should be carried out through the generation of a preliminary on-site report. The second one, a long-term protocol, which provides calculation procedures and constructive solutions for the improvement of the seismic behaviour of affected buildings. The validity of the protocols is demonstrated by specially designed tests, which further illustrate the need for information and communication technologies (ICT) tools in the evaluation of architectonic technical aspects.

The main purpose of the study presented in Cherif et al. [3] was to assess seismic risk and present earthquake loss scenarios for the city of Imzouren, in northern Morocco. The authors resort to an empirical approach to assess the seismic vulnerability of the existing buildings, using the Vulnerability Index Method (RISK-UE), and considering both deterministic and probabilistic earthquake scenarios.

In the same line, Chieffo and Formisano [4] analysed a sub-urban sector of the historic centre of Qualiano, Naples (Italy). The seismic vulnerability of both masonry and reinforced concrete buildings is assessed resorting to a simplified typological-based approach and damage scenarios are created taking into account site and topographical local conditions. The site effects were shown to play an important role in the vulnerability and risk assessment of urban areas.

Estêvão [5] discusses the feasibility of using neural networks to obtain simplified capacity curves for seismic assessment. In the proposed approach, an artificial neural network (ANN) is used by the author to obtain a simplified capacity curve of a building typology, in order to use the N2 method to assess the structural seismic behaviour. The case study presented allowed the conclusion that the ANN precision is very dependent on the amount of data used to train the ANN and demonstrated that it is possible to use ANN to obtain simplified capacity curves for seismic assessment purposes with high precision.

Moving the focus to timber buildings, Drdácký and Urushadze [6] discuss possibilities for seismic improvement of traditional timber carpentry joints. The authors analyse two approaches, each using different retrofitting technologies that avoid completely dismantling the joint, allowing for the conservation of frame integrity. According to experimental observation, fully fastening the brake plates to the wood using screws, proved to be the most effective technique. Reinforcing timber carpentry joints with nails also produced interesting results, as did the use of a combination of nails and inserted plates.

d'Aragona et al. [7], investigate the effect of infills on the lateral seismic capacity of reinforced concrete (RC) framed buildings by explicitly considering possible brittle failures in either unconfined beam-column joints or columns. From a series of non-linear static analyses, the authors observe that the considered existing gravity load designed buildings attain the life safety limit state in a very premature stage of the analysis. They further show evidence that such a poor structural behaviour can be improved through the application of local retrofit interventions whose efficacy, in some cases, varies depending on the consistency of the infills.

Malla et al. [8] discuss the seismic performance of a high-rise RC apartment building with brick infill masonry walls, which was assessed through a nonlinear time history analysis. This work arose in response to the need for more research on the seismic response of this typology, which was identified in a rapid visual assessment campaign performed after the 2015 Gorkha Earthquake. The numerical simulations confirmed that the building sustained reduced damaged for the PGA value of the 2015 Gorkha Earthquake (0.16 g), though for earthquakes with PGA values equal to or higher than 0.36 g, the building would present severe damage, failing to fulfil safety demands.

A discrete element analysis of a shaking table test performed on a traditional stone masonry house is presented by Lemos [9], as a demonstration of the capabilities of this analysis method. Practical application issues are examined, namely the computational requirements for dynamic analysis. Among other remarks, the author stresses that model simplification strategies, whether in terms of geometry or in the constitutive assumptions, are a key to provide meaningful results with the existing data in practical situations.

Finally, Ramírez, Mendes and Lourenço [10] addresses the state of conservation of the Benedictine Monastery of São Miguel de Refojos, located in Cabeceiras de Basto (Portugal), as well as the evaluation of its structural behaviour and seismic performance resorting to nonlinear static analyses. Based on the diagnosis and analyses made, the authors draw a series of conclusions about the need for putting in place a monitoring plan and a set of preventive measures in order to guarantee a more efficient structural behaviour of the building.

The editors would like to express their sincere gratitude to the authors, who have generously shared their scientific knowledge and experience through their contributions, to the peer reviewers, who have significantly contributed to enhance the quality of this Special Issue, and to the managing editors of Building, who have enthusiastically supported and promoted this initiative.

Author Contributions: All authors contributed to every part of the research described in this paper.

Funding: This research was funded by the Portuguese Foundation for Science and Technology (FCT) through the postdoctoral grant SFRH/BPD/122598/2016.

Conflicts of Interest: The authors declare no conflicts of interest.

References

1. Ferreira, T.M.; Mendes, N.; Silva, R. Multiscale Seismic Vulnerability Assessment and Retrofit of Existing Masonry Buildings. *Buildings* **2019**, *9*, 91. [CrossRef]
2. Mascort-Albea, E.J.; Canivell, J.; Jaramillo-Morilla, A.; Romero-Hernández, R.; Ruiz-Jaramillo, J.; Soriano-Cuesta, C. Action protocols for seismic evaluation of structures and damage restoration of residential buildings in Andalusia (Spain): "IT-Sismo" APP. *Buildings* **2019**, *9*, 104. [CrossRef]
3. Cherif, S.; Chourak, M.; Abed, M.; Douiri, A. Potential Seismic Damage Assessment of Residential Buildings in Imzouren City (Northern Morocco). *Buildings* **2018**, *8*, 179. [CrossRef]
4. Chieffo, N.; Formisano, A. The Influence of Geo-Hazard Effects on the Physical Vulnerability Assessment of the Built Heritage: An Application in a District of Naples. *Buildings* **2019**, *9*, 26. [CrossRef]
5. Estêvão, J. Feasibility of Using Neural Networks to Obtain Simplified Capacity Curves for Seismic Assessment. *Buildings* **2018**, *8*, 151. [CrossRef]
6. Drdácký, M.; Urushadze, S. Retrofitting of Imperfect Halved Dovetail Carpentry Joints for Increased Seismic Resistance. *Buildings* **2019**, *9*, 48. [CrossRef]
7. Gaetani d'Aragona, M.; Polese, M.; Di Ludovico, M.; Prota, A. Seismic Vulnerability for RC Infilled Frames: Simplified Evaluation for As-Built and Retrofitted Building Typologies. *Buildings* **2018**, *8*, 137. [CrossRef]
8. Malla, S.; Karanjit, S.; Dangol, P.; Gautam, D. Seismic Performance of High-Rise Condominium Building during the 2015 Gorkha Earthquake Sequence. *Buildings* **2019**, *9*, 36. [CrossRef]
9. Lemos, J.V. Discrete Element Modeling of the Seismic Behavior of Masonry Construction. *Buildings* **2019**, *9*, 43. [CrossRef]
10. Ramírez, R.; Mendes, N.; Lourenço, P.B. Diagnosis and Seismic Behavior Evaluation of the Church of São Miguel de Refojos (Portugal). *Buildings* **2019**, *9*, 138. [CrossRef]

buildings

MDPI

Article

Multiscale Seismic Vulnerability Assessment and Retrofit of Existing Masonry Buildings

Tiago Miguel Ferreira *, Nuno Mendes and Rui Silva

ISISE, Institute of Science and Innovation for Bio-Sustainability (IB-S), Department of Civil Engineering, University of Minho, 4710-057 Braga, Portugal; nunomendes@civil.uminho.pt (N.M.); ruisilva@civil.uminho.pt (R.S.)
* Correspondence: tmferreira@civil.uminho.pt; Tel.: +351-253-510-200

Received: 30 March 2019; Accepted: 17 April 2019; Published: 19 April 2019

Abstract: The growing concern about the protection of built heritage and the sustainability of urban areas has driven the reoccupation of existing masonry buildings, which, in the great majority of the cases, were not designed or constructed to withstand significant seismic forces. This fact, associated with territorial occupation often concentrated in areas with high seismic hazard, makes it essential to look at these buildings from the point of view of the assessment of their seismic vulnerability and retrofitting needs. However, to be effective and efficient, such an assessment must be founded on a solid knowledge of the existing methods and tools, as well as on the criteria that should underlie the selection of the most appropriate to use in each context and situation. Aimed at contributing to systematise that knowledge, this paper presents a comprehensive review of the most relevant vulnerability assessment methods applicable at different scales, as well as the most significant traditional and innovative seismic retrofitting solutions for existing masonry buildings.

Keywords: seismic vulnerability assessment; large-scale vulnerability analysis; numerical modelling; seismic retrofit

1. Introduction

Devastating seismic events keep raising the awareness of scientific, technical and political community to the need of identifying assets at risk and developing more effective and cost-efficient risk mitigation strategies. The capacity to accurately assess the seismic vulnerability of these assets plays a fundamental role in this context, mainly because among the three factors that make up the standard formulation of risk (hazard, vulnerability and exposure), the vulnerability is the only factor in which engineering research can intervene. Nevertheless, and despite the significant advances in the field of seismic engineering and risk assessment, there is still much to be done, particularly with regard to existing buildings, most of them built without antiseismic provisions. Moreover, the wide variety of construction and structural systems, associated with the complex behaviour of their materials (raw earth, timber, masonry, steel and reinforced concrete), greatly limit the application of current codes and building standards to the existing building stock.

To tackle this global issue, it is fundamental to enhance the engagement between innovation and technical stakeholders towards the development and application of more sophisticated and reliable methods of analysis, as well as improved seismic retrofitting techniques compliant with buildings conservation principles, a goal that can only be achieved on the basis of a solid knowledge of the current state of the art. Based on this assumption, the present paper provides a general framework on the issue of the seismic vulnerability assessment of existing masonry buildings by reviewing the most relevant vulnerability assessment methods applicable at different evaluation scales (from large to building scale assessment) and the most successful traditional and innovative solutions used in the retrofitting of existing masonry buildings.

2. Seismic Vulnerability Assessment Approaches at Different Scales

The selection of the most appropriate seismic vulnerability assessment method to use must be made based on a proper balance between simplicity (of the tool) and accuracy (of the results), keeping always in mind that vulnerability outputs can be affected by several sources of uncertainty that should acknowledges and addressed, namely those related to the limitations of the model itself and to the inherent randomness of both the sample and the response.

Over the last decade, several European consortiums have been working on different aspects related to the assessment and mitigation of vulnerability and seismic risk, namely on the classification and categorisation of existing seismic vulnerability assessment approaches at different scales. According to their classification, methodologies can be grouped into three main categories taking into account their level of detail, scale of evaluation and use of data: (1) first level approaches, (2) second level approaches and (3) third level approaches (see Figure 1).

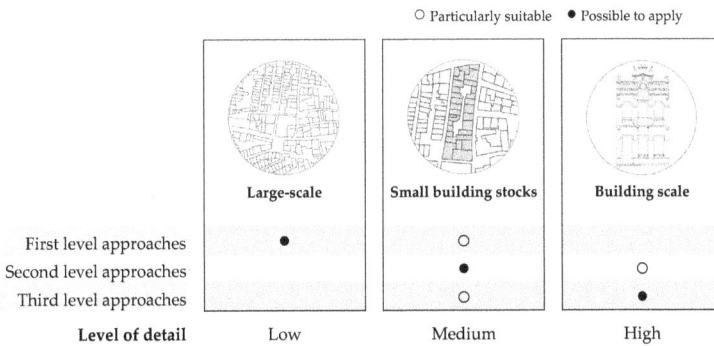

Figure 1. Analysis techniques versus assessment scales.

First level approaches are mainly based on the use of qualitative information and therefore they are ideal for the development of large-scale seismic vulnerability analysis. Second level approaches are based on mechanical models, which rely on detailed geometrical and mechanical information. For this reason, second level approaches are particularly suited to the assessment of small building stocks (aggregates or single buildings). Finally, third level approaches involve the use of complex numerical modelling techniques, which require a comprehensive and rigorous geometrical, material and mechanical characterisation of the building.

3. Large-Scale Seismic Vulnerability Assessment Methods

When assessing the vulnerability of a large number of buildings, the amount of information to collect and treat can be massive and thus the use of more expedite approaches is more adequate. Large-scale seismic vulnerability assessment methods, either at the national or urban level, should be based on few vulnerability indicators, which are usually defined from the statistical treatment of past earthquake damage data. The definition and nature of such methods (some of them are much more quantitative while others are more qualitative) naturally limits their methodological formulation and the level at which the assessment is conducted, from the expedite evaluation of an entire area based on the identification and characterisation of representative building typologies to the individual assessment of each building. Thus, and according to the scheme in Figure 1, some remarkable examples of the application of first and second level approaches are given in the following subsections, categorising them into four different groups of methods: typological, indirect, conventional and hybrid.

3.1. Typological Methods

According to typological methods, buildings are classified into different classes depending on a series of aspects (structural, construction, material, etc.) that either directly or indirectly influence their seismic response. In these methods, vulnerability is described as the probability of a structure to suffer a certain level of damage for a given ground motion intensity level, and damage probabilities are defined on the basis of past earthquake damage data and expert judgement. Since results obtained using typological methods are based on field investigation, they have to be considered in terms of their statistical accuracy. In this sense, they are valid only for either that particular area or for other areas with similar field conditions (in terms building typology and seismic hazard). A striking example of the use of a typological approach are the Damage Probability Matrices (DPMs) developed by Whitman et al. [1]. In this work, the authors have compiled DPMs for several building typologies according to the damaged sustained in over 1600 buildings after the 1971 San Fernando earthquake. According to different authors [2,3], one of the first European versions of a DPM was produced in Italy by Braga et al. [4] from the statistical treatment of the damage data collected after the 1980 Irpinia earthquake. The authors have used a binomial distribution to describe the damage distributions of each class for different seismic intensities. Buildings were separated into three vulnerability classes (A, B and C) and a DPM based on the MSK scale [5] was evaluated for each class [6]. The Italian National Seismic Service has processed the same Irpinia 1980 database, in order to obtain DPMs, see Di Pasquale et al. [7]. The main differences between this work and the original DPMs proposed by Braga et al [4] lies in the use of dwellings instead of buildings and in the quantification of the earthquake intensity in terms of MCS scale [8], instead of the MSK scale [5].

Although, as already referred, its origin goes back to the 1970s, the use of DPMs is still very popular. In 2003, as part of the ENSeRVES project (European Network on Seismic Risk, Vulnerability and Earthquake Scenarios), Dolce et al. [9] have derived DPMs for the Italian town of Potenza. An additional vulnerability class (D) was included by the authors in the formulation using the EMS-98 scale [10] in order to account for the buildings constructed after 1980 (i.e., buildings that have either been retrofitted or designed to comply with recent seismic codes) [2,6]. More recently, Lagomarsino and Giovinazzi [11] proposed a set of DPMs from the European Macroseismic Scale (EMS) aimed at providing a model for estimating seismic damages in five increasing and perfectly defined levels of damage.

3.2. Indirect Methods

Indirect methods involve the determination of a vulnerability index and the establishment of a series of relationships between damage and seismic intensity [6], which are generally supported by statistical studies of postearthquake damage data. As schematised in Figure 1, indirect methods have been applied extensively to assess the seismic vulnerability of large areas and/or building stocks. The "Vulnerability Index Method", originally proposed by Benedetti and Petrini [12] (see also [13]), is one of the most applied indirect methods, involving the computation of a vulnerability index (i.e., a building vulnerability classification system), which aims at traducing the physical and structural characteristics of the building into a quantitative form. According to this approach, each building is classified in terms of a vulnerability index related to a damage grade determined via the use of vulnerability functions. These functions enable the formulation of the damage suffered by buildings for each level of seismic intensity (or peak ground acceleration, PGA) and vulnerability index, see Figure 2. Indirect methods use extensive databases of building characteristics (typological and mechanical properties) and rely on observed damage after recent earthquakes to classify vulnerability, based on a score assignment. An adapted vulnerability index method has been developed and used to assess the seismic vulnerability of both unreinforced masonry and reinforced concrete building under the scope of "Catania Project" [14]. Following the guidelines of ATC-21 [15], a rapid screening approach was used in this project to define the vulnerability scores of the buildings.

Among the main advantages of indirect techniques, it is worth highlighting their ability to determine the vulnerability characteristics of the building stock under consideration, rather than base the vulnerability definition of a certain building typology considered as representative of the that building stock. Nonetheless, since they are based on vulnerability indices that are defined on the basis of empirical-based coefficients (weights), the uncertainty associated to the results is potentially high and should therefore be taken into account. Moreover, large-scale assessment using indirect techniques (at a regional or a national scale) must be based on data gathered from a large number of buildings, assumed to be representative of the building stock [16].

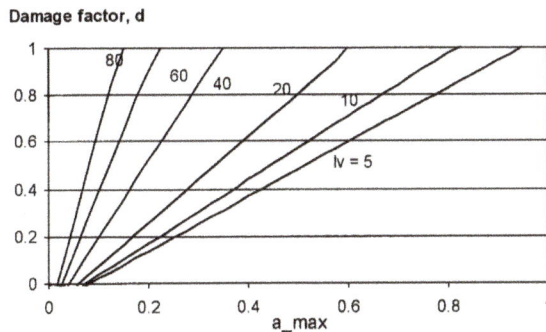

Figure 2. Unified relationships between vulnerability, damage and peak ground acceleration, adapted from Petrini [17].

3.3. Conventional Methods

Conventional techniques are essentially heuristic, proposing a correspondence between a vulnerability index and a certain level of damage. According to Vicente et al. [18], there are essentially two groups of conventional approaches: the group that qualify the different physical characteristics of structures empirically and the group of those based on the criteria defined in seismic design standards for structures, evaluating the capacity–demand relationship of buildings. One of the best-known approaches of the first group is probably ATC-13 [19]. Damage probability matrices were defined therein based on the know-how of more than 50 senior earthquake engineering experts who were asked to estimate damage factors (the ratio of loss to replacement cost, expressed as a percentage) for Modified Mercalli Intensities [20] from VI to XII, and for 36 different building classes. Some large-scale risk and loss assessment works based on ATC-13 can be found in the literature [21–25]. As an example, Table 1 presents the damage factors for each damage state, computed by Eleftheriadou and Karabinis [22] for the city of Basel.

Table 1. Damage probability matrix (DPM) presented by Eleftheriadou and Karabinis [23] for Basel.

Damage Grade	CDF (%)	Intensity					
		V	VI	VII	VIII	IX	X
1	5	8	35	20	20	5	0
2	20	0	8	37	37	15	7
3	55	0	0	35	35	37	15
4	90	0	0	8	8	35	37
5	100	0	0	0	0	8	41
MDF		0.4	3	14	35	63	84

3.4. Hybrid Methods

As its name suggests, the hybrid techniques combine features of the above-mentioned techniques. A representative example of a hybrid technique is the macroseismic approach proposed by Faccioli et al. [11], which combines the characteristics of the typological and indirect methods using the referred vulnerability classes and a vulnerability index improved by the use of modifier parameters. This macroseismic method, either in its original format or with some minor adaptations, has been widely applied for assessing the seismic risk of various urban areas, such as Barcelona (Spain) [21], Faro (Portugal) [26] or Annaba (Algeria) [27], just to mention some, as well as of neighbourhoods [28] and building aggregates [29]. An interesting example of the combined application of a hybrid approach and a GIS tool was recently presented by Ferreira et al. [30]. In that work, the authors make use of a simplified index-based approach to evaluate the impact resulting from the adoption of different large-scale retrofitting strategies, measured in terms of material, human and economic losses. In order to improve their interpretation, the results are mapped using a commercial GIS software, wherein georeferenced graphical information was combined and connected to a relational database containing the main structural characteristics of the buildings, see Figure 3. Various modules with different objectives were developed and integrated into the GIS, including vulnerability assessment, damage and loss estimation (number of collapsed buildings, casualty rate, number of unusable buildings and repair costs) for different macroseismic intensities, allowing for the construction of multiple physical damage and loss scenarios. With this discussion, the authors intended not only to demonstrate how simplified seismic vulnerability assessment approaches can be used to analyse the impacts resulting from the implementation of large-scale retrofitting programs (Figure 3a versus Figure 3b), but also to prove that investing in prevention strategies designed to mitigate urban vulnerability is one of the most effective strategies, both from the social and economic standpoint.

(a)   (b)

Figure 3. Mapping of damage scenarios for the old city centre of Horta: (**a**) before and (**b**) after the application of seismic retrofitting strategies, adapted from Ferreira et al. [30].

In the same line, Aguado et al. [31] took advantage of a hybrid approach to discuss the potential benefit resulting from the application of different seismic retrofitting strategies, not only accounting for their effect on the reduction of individual and global damages resulting from different seismic scenarios, but also regarding their global impact in terms of civil protection and urban accessibility. For such, the authors applied a hybrid seismic vulnerability assessment approach for façade walls to obtain vulnerability and damage scenarios for the historical city centre of Coimbra. Resorting to a GIS tool, these scenarios were subsequently used to map evacuation routes with different levels of accessibility, see Figure 4a. Blocked, restricted access and free roads were plotted considering both the façade walls prone to partial or total collapse and the narrowness of streets. Moreover, the urban areas that may be potentially inaccessible in the case of a seismic event were also identified, as well as the number of people that, in consequence, may be affected. Finally, the author has analysed the impact, in terms of urban accessibility, of retrofitting the building façade walls that define the areas more prone inaccessibility, see Figure 4b. Due to its immediate and practical application, it is easy to understand that this kind of information is of major significance for decision-makers and civil protection agencies.

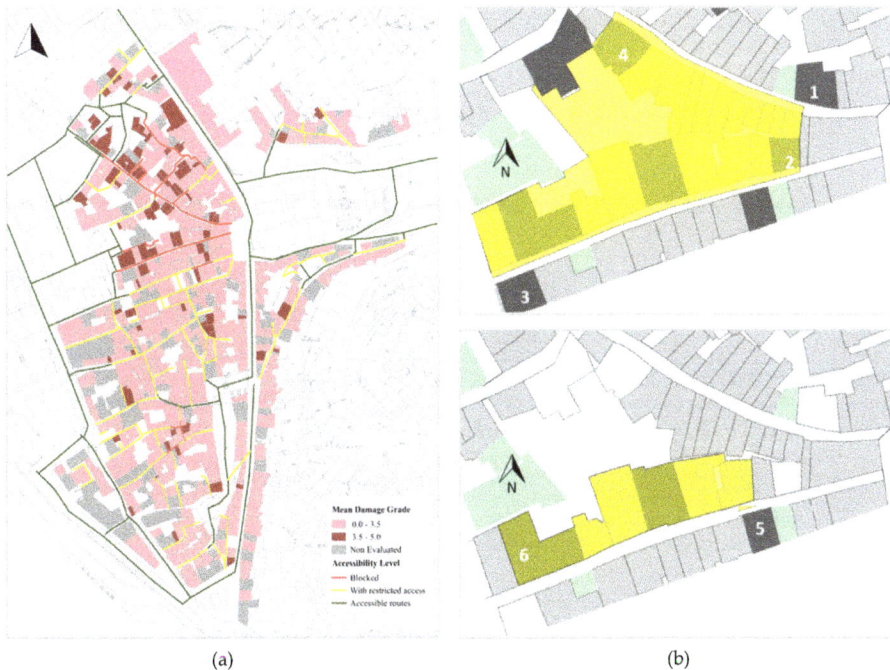

(a)   (b)

Figure 4. Mapping of (**a**) evacuation routes and (**b**) configuration of the most critical urban area before (top) and after (bottom) retrofit, adapted from Aguado et al. [31].

4. Seismic Vulnerability Assessment at Building Scale

The seismic behaviour of existing masonry buildings is complex and depends on several aspects. The evaluation of the seismic response of this type of buildings can be divided in two main types of behaviour: (1) in-plane behaviour and (2) out-of-plane behaviour. In the last decades, several experimental and numerical studies on the in-plane and out-of-plane of existing masonry buildings were carried out, which allowed for significant improvement of the knowledge on its seismic performance. Furthermore, these studies helped develop new tools for the seismic assessment of existing masonry buildings. In the next sections, an overview on the numerical modelling approaches at material level, and the numerical modelling methods, the types of analysis and the assessment criteria for the seismic assessment of existing masonry structures at the building scale are presented.

4.1. Numerical Modelling Approaches at Material Level

Masonry is a composite material with units and joints. The units can be adobe bricks, fired bricks, ashlars, irregular stones and others. Two main types of joints can be distinguished, namely, dry joints and mortar joints. The mortar joints can be such as mud mortar, lime-based mortar and cement-based mortar, among others. The structural behaviour of masonry depends on several aspects: (a) material properties of units and joints; (b) geometrical properties of units and joints; (c) quality of the workmanship; and (d) conservation status. The morphology of the cross-section of the masonry element has also influence on the structural, since the masonry can present: (a) different types of unit bonds, such as uncoursed bond and coursed bond with stone units or the English bond and Flemish bond with brick units; (b) different number of layers/leafs that compose the masonry element (single leaf masonry wall, multileaf masonry wall and multilayered vault); and (c) different types of arrangement at the connections between the walls (with and without interlocking).

The strategy for the numerical modelling of existing masonry buildings can be defined based on the assumptions assumed at material level. At material level, three elements can be distinguished in the masonry with mortar (Figure 5a): units, mortar and interface unit/mortar. Two main approaches can be adopted to simulate masonry [32]: micro- and macromodelling. Micromodelling can be further divided into [32] detailed micromodelling and simplified micromodelling. In the macromodelling approach (Figure 5d), the units, the mortar and the interfaces unit/mortar are modelled together as a continuum and composite material. In the simplified micromodelling approach (Figure 5c), the units are expanded (unit and half of the thickness of the mortar) and they are simulated as continuum elements. The behaviour of the mortar and of the interface unit/mortar is modelled by discontinuous elements. The third approach (Figure 5b) corresponds to the most detailed approach, in which the units and the mortar are simulated by independent continuum elements and the behaviour of the interfaces unit/mortar are modelled by discontinuous elements. The mortar with dry joints can be simulated using a micromodelling approach, composed by continuous elements for the units and discontinuous elements to simulate the interfaces unit/mortar (infinite compressive stiffness and strength and no tensile stiffness and strength). In general, the most detailed modelling approaches provide simulations that are more accurate and can be more representative of the real behaviour. However, they are more complex, since they require a high number of material parameters and more computational effort. The number of material parameters is significantly high when the nonlinear analysis is adopted. Thus, the selection of the modelling approach at material level should be based on an equilibrium between the required precision in the results and the effort needed to prepare the model, run the analysis and analyse the results. In general, macromodelling is used to simulate entire buildings, while micromodelling is adopted to model small portions/elements of masonry. For examples of applications of the macromodelling, simplified micromodelling and detailed micromodelling see respectively Mendes and Lourenço, Kurdo et al. and Andreotti et al. [33–35].

4.2. Numerical Modelling Methods

In the numerical modelling of existing masonry buildings, four main strategies based on the following methods can be distinguished: (1) Finite Element Method (FEM); (2) Distinct Element Method (DEM); (3) equivalent frame method; and (4) method using rigid macroblocks taking into account the expected collapse mechanisms.

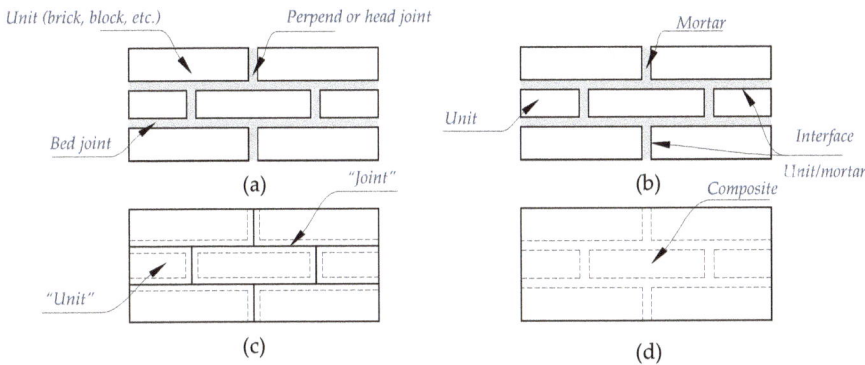

Figure 5. Modelling strategies at material level: (**a**) masonry sample; (**b**) detailed micromodelling; (**c**) simplified micromodelling; and (**d**) macromodelling, adapted from Lourenço [32].

In the Finite Element Method (FEM) (Figure 6b), the continuous domain is discretised by a mesh composed by elements connected by nodes. It is a method widely used to solve complex engineering problems, in which the problem is described by differential equations. In structural problems, first a numerical model, composed by elements, is prepared. Then, and after running the analysis, the results are obtained and evaluated. The FEM has several types of 1D, 2D and 3D elements based on different formulations that can be adopted to simulate structural problems, such as beams elements, shell elements or solid elements. FEM has also available other types of elements to simulate particular behaviours/characteristics of the structure, such as interface elements, springs elements, mass elements, boundary elements or reinforcement elements. This method can be used to solve static and dynamic problems, with linear and nonlinear behaviour. The nonlinear analyses can be adopted to simulate the nonlinear behaviour of the materials (e.g., cracking and crushing of masonry) or the geometric nonlinearity. The FEM models can consider also the anisotropic behaviour of the masonry. The standard FEM method to solve nonlinear problems is implicit, which requires an iterative method, more computational effort and provides an unconditionally stable solution. There are also FEM codes that provide the explicit method, aiming at reducing the computational effort associated to the convergence difficulties of the implicit method. However, this FEM method is unconditionally stable and small steps should be adopted. Several softwares to solve complex structural problems, including the behaviour of existing masonry buildings, are available, such as DIANA^{TM} [36], ANSYS® [37] or Abaqus Unified FEA^{TM} [38].

The FEM has been widely used to evaluate the structural performance of existing masonry buildings, including for the seismic action. Most of the FEM models of entire buildings are developed based on the macromodelling approach (Section 4.1), even for historic buildings, since it is the most simple and fast approach to prepare the models and run the analysis. In general, this approach is a good compromise between required accuracy and computational effort to evaluate the global seismic behaviour of an entire masonry building. There are several applications of the macromodelling approach based on the FEM to evaluate the performance of masonry buildings, taking into account the in-plane and out-of-plane behaviour of the walls, and to evaluate the efficiency of strengthening techniques [39–41]. The numerical models developed in the FEM based on the micromodelling approach (Section 4.1) are detailed, requiring more computational effort. Thus, this modelling strategy is not commonly adopted simulate the seismic behaviour of entire masonry buildings. It is mainly adopted to simulate a small portion of masonry or a masonry element, such a masonry wall (e.g., [42,43]).

The Distinct Element Method (DEM) (Figure 6a) was mainly developed to solve problems related to discontinuous materials, such as soils or masonry. Due to high computational effort required to run DEM models, this method was used mainly to simulate behaviour of small models. Nowadays, and to the recent advances in the computer hardware, the DEM is becoming more popular, even to solve problems of civil engineering, such the evaluation of the seismic performance of masonry structures. In the analysis of masonry structures or masonry components, two approaches can be adopted for DEM models: (1) DEM models developed with rigid or deformable blocks and (2) DEM models developed with spherical particles. The DEM models with spherical particles involve a high number of elements and they are not appropriated to simulate entire masonry buildings. In general, the masonry models based on DEM are prepared with blocks to simulate the units, in which the interaction between the blocks is defined by contact points. The normal and shear stress at the contact points are based on the relative displacement caused by applied forces. In this method, the accuracy is guaranteed if an enough number of contact points are adopted [44]. When the deformable blocks are adopted, the blocks are discretised into an internal finite element mesh (combined FEM/DEM method [45]). The DEM allows stimulating the collapse mechanisms at large displacement domains, which is useful to evaluate the behaviour of masonry buildings from the initial damage until near its collapse, including damage caused by the sliding mechanism. The solution is obtained from an explicit time-stepping algorithm, which allows to simulate the nonlinear interaction of a large number of blocks without high computational requirements or to use an iterative method. Thus, small steps should be used. The UDEC [46], 3DEC [47] and DEMpack [48] are examples of software based on DEM. In general, these softwares have also beam and cables elements to simulate strengthening of the structure.

The DEM can be useful to take into account the effect of the morphology of the masonry, using one block to simulate each unit, which require a high computational effort for large models. The computational effort can be reduced if the blocks simulate portions of masonry composed by several units. In this simplification, the discretisation should be carefully defined, taking into account the expected collapse mechanisms, since it can have influence on the response of the model. There are several applications of the DEM to evaluate the seismic performance of existing masonry structures, mainly for historic constructions [49–51].

In the equivalent frame method (Figure 6c), the main objective is to represent each type of member of the masonry building (piers, lintels, spandrels, etc.) by a specific structural element that simulate its behaviour (1D or 2D FEM element, macroblock, macroelements, etc.). In general, these structural elements are connected by rigid nodes, springs or interface elements. Thus, a 3D equivalent frame to the structural system of the buildings is created. This type of modelling strategy started to be applied by Tomaževič [52] in 1978, where the 1D macroelements were adopted to simulate the diagonal shear failure of the piers and assuming that the in-plane global behaviour of masonry walls is mainly controlled by this type of failure (POR method). In the last decades, new formulations based on this modelling strategy have been developed, aiming at including other types of behaviour and failures with influence in the global seismic behaviour of masonry buildings, such as both flexural and shear behaviour of piers and spandrels or the sliding failure of the piers. The analysis based on this numerical modelling strategy is significantly faster than the analysis carried out using the previous methods, since involves a very low number of degrees of freedom. Several software using the equivalent frame method are available for masonry buildings, such as 3Muri [53], PRO_SAM [54] and 3DMacro [55]. In general, this software allows users to take into account the contribution of other types of structural elements, such as timber floors or reinforced concrete floors. The analysis of existing masonry structures based on the equivalent frame methods can be also performed in software's that were not originally developed for masonry structures and more common in design offices, such SAP2000® [56], simulating the piers and the columns through beam elements with vertical and/or horizontal rigid offsets, connected by plastic hinges (spring elements to simulate the shear and the flexural behaviour) [57]. For examples of applications of this modelling strategy to existing masonry buildings see Lagomarsino et al., Magenes et al. and Penna et al. [58–60].

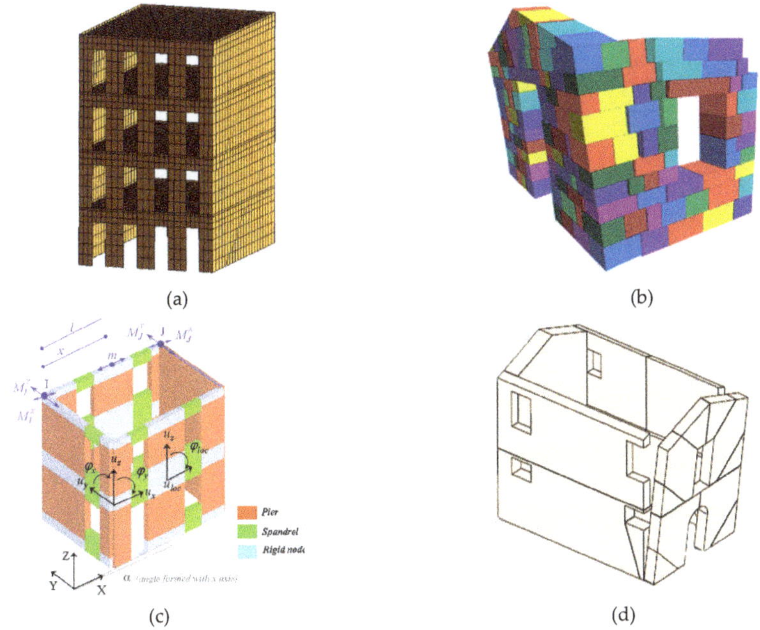

Figure 6. Examples of numerical model of masonry buildings using: (**a**) FEM [61], (**b**) DEM [50], (**c**) simplified equivalent frame method [62] and (**d**) method with rigid macroblocks taking into account the expected collapse mechanisms [63].

Finally, in last numerical modelling strategy (Figure 6d), the masonry building is discretised into several rigid macroblocks taking into account the expected collapse mechanisms. In general, and contrarily to the previous method of modelling, in this modelling strategy the macroblocks are defined by the failure lines and are larger than the structural components of the masonry walls, which reduces the number of elements and computational effort. The expected collapse mechanisms can be defined based on damage observed in postearthquake surveys, results obtained from experimental tests, numerical results obtained from analysis with FEM and DEM models and practitioner experience [64]. This means that the assumptions adopted for the discretisation of the structure can have influence on the evaluation of its seismic response, mainly when the collapse mechanism associated to the lowest load capacity is not considered. In general, the classic limit analysis (no tensile strength along the macroblock interfaces, infinite compressive strength for the macroblocks and the sliding failure is not considered) is used in modelling strategy, which makes it a practical and fast numerical tool to evaluate the seismic behaviour of existing masonry buildings. However, more advanced limit analysis with rigid blocks has been developed, for example, considering the crushing of the masonry and the sliding failure [65,66]. This numerical modelling strategy can be also adopted to evaluate easily the seismic capacity of local failures of masonry buildings using the static equilibrium equations (e.g., overturning of the façade with rotation at the base of the wall). For examples of application of this modelling strategy to masonry structures see Orduña [63] and Mendes [67].

4.3. Types of Analysis

The previous numerical modelling strategies can be adopted to perform different types of analysis to obtain the seismic response of existing masonry buildings. Five main types of structural analysis are highlighted: (1) limit analysis; (2) linear static analysis; (3) nonlinear static analysis (pushover analysis); (4) linear dynamic analysis; and (5) nonlinear dynamic analysis with time integration.

In general, the limit analysis, which can be based on the static or the kinematic approach, requires a low number of material properties and low computational effort. It provides results on the collapse mechanism and the load capacity of the structure. This type of analysis is a very useful tool for the evaluation of the seismic performance of existing masonry buildings [67].

The linear static analysis is mainly used to design new structures and, since does not considered the nonlinear behaviour of the masonry, it is not an analysis appropriated to evaluate the seismic behaviour of masonry structures near the collapse. This limitation is extended to the linear dynamic analysis, such as the modal analysis or the linear dynamic analysis with time integration.

In the evaluation of the seismic performance of existing masonry buildings, the nonlinear static analysis (pushover analysis) is the most used. In this type of analysis, the nonlinear behaviour of the masonry (e.g., cracking and crushing) is considered, and the seismic action is applied by using horizontal static loads. The horizontal loads can be proportional to the mass or to the shape of the fundamental mode of the structure and they are increased until the collapse of the building. In general, it requires the applications of several nonlinear parameters of the masonry and an iterative method to obtain the solution (implicit analysis), which can cause some convergence difficulties. Besides these aspects, it is a powerful tool to simulate the behaviour of masonry and a good compromise between accuracy of the results and effort computational. It provides results fundamental to evaluate the response of masonry buildings, such the capacity curve (applied force vs. displacement at the control point), where it is possible to know the inception of the nonlinear behaviour, the maximum capacity of the structure and zones of concentration of damage. Several procedures using the pushover analysis have been developed to estimate the seismic response of structures, such as the N2 method [68], the capacity spectrum method [69] or the displacement coefficient method [70]. It should be noted that the first nonlinear static approaches were mainly developed for regular concrete or steel structures with rigid diaphragms. Thus, more recent and advanced nonlinear static approaches have been developed or adapted for existing masonry structures, for example considering the effects of the in-plan and in-elevation irregularly of the building [71].

Finally, the nonlinear dynamic analysis with time integration is the most advanced type of analysis, in which accelerograms are applied at the base of the structure. However, it is not practical for the evaluation of the seismic behaviour of existing masonry building, since requires the applications of several nonlinear material properties of the masonry, including the damping ratios and advanced mathematical tools to solve the equation of motion at each time step. In general, the FEM uses the implicit solution with an iterative method by default and the solution for DEM is obtained using the explicit method. Although this type of analysis provides detailed and precise results on the dynamic response, it requires advanced knowledge on structural analysis, interpretation of results and high computational effort. For examples of nonlinear static and dynamic analyses carried for the evaluation of the seismic performance of masonry buildings see Mendes and Lourenço [61] and Chácara et al. [72].

4.4. Assessment Criteria

After the selection of the numerical modelling strategy to simulate the structure, the selection of the type of analysis, to obtain the numerical response of the building, and the seismic assessment can be carried out, comparing the capacity of the structure with the demand criteria. Three main approaches for the seismic assessment can be adopted: (1) Force-Based Approach (FBA), (2) Displacement-Based Approach (DBA) and (3) Energy-Based Approach (EBA).

In the FBA, the internal forces (e.g., axial forces, shear forces or bending moments) are compared with the demand forces defined in the codes for the seismic analysis of structures. This type of approach is widely used to design new structures, for the ultimate and service limit states. In the seismic assessment of structures based on the FBA and using the typical linear static analysis for design, a behaviour factor is adopted to take into account the inelastic cyclic deformation and the energy dissipation capacity of the structural system. However, the deformation is better related to the severity of the damage and the dynamic behaviour near the collapse. Thus, the DBA seems more appropriated to evaluate the seismic response of existing masonry buildings, mainly to evaluate the rocking mechanism. In this approach, the response in terms of deformation (e.g., displacements or drifts) is compared with the limit states defined in the seismic codes/literature, which are associated to the severity of the damage (e.g., no damage, minor damage, severe damage and collapse) [72]. Finally, the seismic assessment of a structure can be performed comparing the demand energy with the required energy to obtain the collapse of the building (EBA) [73]. Although this assessment approach is not commonly used, some research on the overturning of masonry walls based on the EBA has been performed [74].

5. Seismic Retrofitting Techniques for Masonry Buildings

When a building system or an existing structure are deemed unsafe under seismic loads or present evident limitations regarding these actions, improvement of the seismic behaviour is required to reduce the seismic vulnerability and to allow safe use. The awareness of the humankind to seismic risk occurred centuries ago and led to the development of several traditional seismic resistant building practices, which aimed not only at improving the local building processes, but also the existing buildings. The development of these practices was based on the empirical knowledge gained through time by the local populations struggling against the repeating destruction of several earthquakes. The most successful traditional practices are nowadays identified in many buildings around the world and are recognised as evidences of local seismic cultures [75]. On the other hand, the recent developments on earthquake engineering made the building practices and strengthening strategies become supported by a scientific background. These recent developments allowed also the proposal of innovative seismic retrofitting strategies based on advanced technologies and materials, which grant high structural efficiency.

In general, the traditional retrofitting solutions follow five main principles: (1) improvement of the connections between structural elements to allow a better global behaviour; (2) stiffening of the horizontal structural elements; (3) consolidation of the vertical structural elements; (4) use of redundant structural elements as backup for possible partial failure; and (5) construction of additional structural elements contributing to the resistance of the building to horizontal loads.

The innovative retrofitting solutions added two more principles: (1) improvement of the mechanical properties of the masonry walls to increase the loading and displacement capacities and (2) reduction of the dynamic effects induced by earthquakes.

The following sections present a brief discussion on the most successful traditional and innovative solutions used in the retrofitting of existing masonry buildings.

5.1. Traditional Solutions

Very frequently, ancient masonry buildings present timber floors and roofs, which are typically insufficiently stiff to provide to the structure the so-called "box behaviour". "Box behaviour" is responsible for promoting the structural collabouration between the masonry walls, meaning that in its absence the global seismic response of a building is considerably limited [39]. Furthermore, these buildings are also characterised by poor floor–wall, roof–wall and wall–wall connections, which do not allow the structure to develop a satisfactory monolithic "box behaviour" [76]. Thus, it is compressible that many of the traditional strengthening solutions are focused in the improvement of these limitations of ancient masonry buildings.

The introduction of ring beams between the walls and the roof is one of the most effective solutions to ensure the structural integrity of masonry buildings. Traditionally, ring beams are built using a pair of longitudinal planks joined together by transversal elements in a ladder-like configuration [77]. Nevertheless, they can be built using more rough timberworks, such as a grid of horizontal timber trunks or tree branches lying longitudinally and transversally [78]. In the last few decades, ring beams were also built using reinforced concrete, but this procedure introduces compatibility concerns, whose discussion is beyond the scope of this paper. The introduction of ring beams can lead to different types of improvement of the structural behaviour, namely enhancement of the "box behaviour" (increase of the loading and deformation capacity), prevention of out-of-plane collapse of the walls and tying of masonry leaves in multileaf walls. Thus, ring beams should be continuous around the entire building and present adequate connection at the intersections between walls and between the ring beam and the walls in order to provide the best performance. The main drawback of ring beams is its installation, since it may require removing and raising the roof.

Ties have been also extensively used for seismic strengthening of ancient buildings and are typically composed of timber tie beams or steel rods connecting walls in opposing façades (see Figure 7a). In this last case, steel anchor plates are used to fasten conveniently the tie rod to the masonry, meaning that this solution only works when the walls tend to move away from each other. Nevertheless, it enhances the "box behaviour" of the building by avoiding the out-of-plane collapse of the façades and by constraining the floors and roof. Thus, steel ties are typically introduced at the floor and roof levels, thus avoiding interference with the habitable spaces. The dimension of the anchor plates is another important aspect of this strengthening solution, as they should be sufficient large to avoid shear failure of the masonry. It should be noted that in buildings with vaulted masonry floors and roofs, ties are applied with similar purpose, namely to limit relative lateral movements of the abutment walls caused by seismic loads. This type of strengthening promotes the preservation of the geometry of the vault, avoiding its collapse.

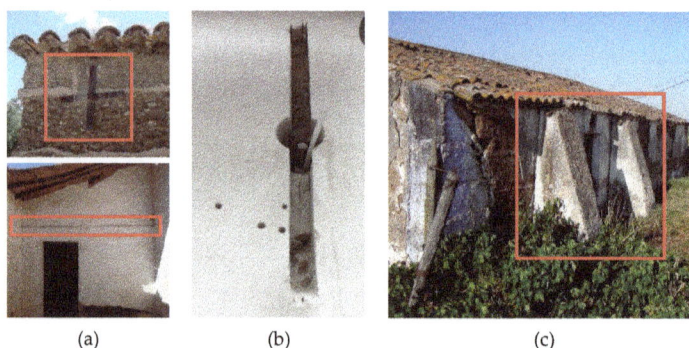

(a)   (b)   (c)

Figure 7. Examples of traditional strengthening solutions: (**a**) steel tie, (**b**) installation of an anchor system and (**c**) buttresses.

The strengthening of connections of timber roofs and floors with the masonry walls is traditionally achieved with the use of timber wedges, which ensure a tight mechanical connection between the joists and the walls. In case of absence of this solution in existing buildings, its implementation requires sufficiently long resting lengths of the joists. In alternative, metal brackets, steel straps and ties can be more easily implemented to connect the timber floors and roofs to the walls [79]. Partial reconstruction can also be used to introduce or substitute timber wall plates, which ensure a better transition of the roof or floor timber elements to the walls. As referred previously, proper roof–wall and floor–wall connections are fundamental to promote the "box behaviour" of the structure.

The strengthening of masonry buildings by means of the improvement of the connection wall–wall is also frequently observed in existing masonry, namely at the corners. This type of improvement has been mainly implemented during the construction of these building by employing materials with better mechanical properties at these sections, which include high quality stones, bricks and timber elements. In existing buildings without these solutions, their implementation would require a large amount of reconstruction work, nevertheless the bond between original and new materials can be questionable. In this case, the strengthening of the wall–wall connection is more easily implemented by using anchor systems (Figure 7b), where a rebar is introduced in the masonry and is effectively bonded by means of grout injection and an anchor plate [80]. The additional tensile strength provided by this strengthening solution improves the load transmission capacity between intersecting walls, enhancing the "box behaviour".

Stiffening of floors and roofs made of timber is an effective solution to improve their diaphragmatic behaviour and, consequently, to improve the "box behaviour" of the existing structure. Floors and roofs can be stiffened by means of diagonal bracing, triangulation and by providing an additional layer of sheathing boards or timber planks placed perpendicularly to the existing ones [81]. Stiffening can also be achieved by substituting the timber floors with slabs made of reinforced concrete or mixed concrete clay blocks, which require building reinforced concrete beams on the walls to conveniently support the slabs. These beams should grant an adequate connection with the walls and provide uniform load transmission [82]. Nevertheless, this solution is highly intrusive, meaning that its use may be unacceptable.

Consolidation of the masonry walls (vertical structural elements) is a particularly used in multileaf walls. Under seismic loading, the masonry leaves tend to behave independently from each other if they are not properly connected, meaning that they are more susceptible to separation and thus to out-of-plane failure. The connection between leaves is typically assured by through-stones, which consist of long stones placed through the full thickness of the wall. These elements serve as connectors between leaves and allow an enhanced load distribution [51]. In multileaf walls without these elements, proper connection can be achieved by means of transversal ties, where a rod is introduced through the full thickness of the wall and is properly fixed with an anchor system and/or grout injection [83]. Consolidation of masonry walls also involves repairing existing cracks (e.g., postearthquake cracks) that debilitate the in-plane and out-of-plane performance of the walls. Cracks can be repaired by means of partial reconstruction, soft stitching and stapling. The first technique is very labour-intensive and intrusive, meaning that its use is only justified in the case of severe damage. Soft stitching consists in opening chases crossing the cracks, which are then filled with masonry units to create a mechanical bond. Stapling consists in applying staple shaped steel rebars transversally to the crack to connect the split sides [84]. Repointing (or deep repointing) is another technique employed to consolidate masonry and consists in removing the bed–joint mortar to substitute it by a better one. This procedure improves the accommodation between units and the bond in the masonry. In some situations, reinforcing rebars are also introduced with the mortar to mitigate other structural problems, such as creep damage [85].

Openings are clearly weak points of masonry walls, which must be properly designed to avoid collapse during seismic excitation. In this regard, lintels are key elements that must grant proper redistribution capacity of the transmitted loads. Lintels can be constituted by single stones or timber elements with length longer than the span and/or by discharging masonry arches. Thus, strengthening can be achieved by reinforcing the lintel resisting elements or by substituting them by stronger ones. Another option to reduce the seismic vulnerability associated with openings is to reduce their span or, as a more drastic solution, close them permanently [75].

Providing redundant structural elements to an existing building allows avoiding full collapse, even if partial collapse of the most vulnerable structural elements occurs [86]. Thus, this strategy is not efficient to mitigate damage. Rather than that, it allows to preserve the life of inhabitants by avoiding falling parts of the building to the interior. To this purpose, an additional structure made of timber, reinforced concrete or steel is built inside the existing building, which nowadays may constitute an intervention excessively invasive and likely unacceptable.

The construction of buttresses is a strengthening solution that serves to help the walls of a building counteracting the horizontal loads induced by earthquakes. Buttresses are massive masonry walls with typical triangular shape (see Figure 7c), which limit the out-of-plane movement of the adjacent walls due to the contribution of their heavy weight and additional shear stiffness and strength. Thus, buttresses and adjacent walls should be properly connected with cross ties in order to avoid an independent behaviour when both move apart, as undesirable impact damage may occur [78].

5.2. Innovative Solutions

The development of innovative seismic strengthening solutions originated new approaches for interventions in existing masonry buildings. Despite structural strengthening efficiency being the main motivation associated with these solutions, in their initial proposal, not all cases respected the main principles established in the conservation charters. Nevertheless, further experience and scientific studies helped improve many aspects of some of these solutions, among which the fulfilment of compatibility requirements was a main focus. The following paragraphs present a brief discussion of the main innovative strengthening solutions developed for existing masonry buildings, namely grout injection, externally bonded composites, base isolation and energy dissipation.

The strengthening of multileaf masonry walls with grout injection consists in filling their inner voids and cracks with a fluid mortar (see Figure 8a), which after hardening provides continuity and bond to the material [87]. Thus, the connection between leaves, and the overall stiffness and strength of the walls are significantly increased [83], leading to an improved seismic behaviour [88]. The capacity of grouts to penetrate into small pores also enables this solution to be used in the repair of cracks [89]. Despite the advantages of grout injection, it presents important drawbacks, such as the unpredictability of the required grout volume and the irreversibility of the solution. This unpredictability compromises the cost control of the intervention due to the impossibility of measuring adequately the ratio and interconnectivity of the voids, which, in general, represent an important percentage of the total wall volume (as high as 30%). Thus, the complete injection of the walls in the building can be a costly operation, meaning that its use in large buildings may be prescribed only for the most vulnerable walls. The irreversibility of grout injection also raises concerns on the selection of the grout, thus the use of poor compatibility grouts should be avoided. In this regard, past interventions evidenced that epoxy- and cement-based grouts may introduce durability problems in ancient masonry walls. Nowadays, several commercial grouts specifically developed for this type of masonry are available and are mainly composed of lime and pozzolanas, which provide enhanced compatibility.

(a)   (b)

Figure 8. Examples of innovative strengthening solutions: (**a**) grout injection and (**b**) textile-reinforced mortar (TRM).

The strengthening of masonry with externally bonded composites has been also used to reduce the seismic vulnerability of masonry buildings, especially in Italy, where these solutions were mainly based on fibre-reinforced polymers (FRP). The popularity of FRP-based strengthening was driven by its ease of application and high efficiency in increasing the tensile strength and ductility of masonry walls, while increasing negligibly the weight (high strength and stiffness to weight ratio). Nevertheless, it presents several drawbacks related with the use of organic matrixes (epoxy-based), such as poor fire resistance, lack of vapour permeability, low reversibility, high-cost and poor compatibility with masonry substrates [90,91]. Thus, alternative composite solutions have been developed by integrating matrixes with enhanced compatibility, namely cement-, lime- or clay-based mortars. The adequate embedding and bond of these mortars also required substituting the sheet textiles by mesh textiles or steel grids with adequate aperture sizes (Figure 8b). These alternative composites are known as Steel Reinforced Grout (SRG), Fibre-Reinforced Cementitious Matrix (FRCM) or Textile-Reinforced Mortar (TRM) [92,93]. Nowadays, several commercial TRM systems are available in the market, as provided by companies such as Ardea Progetti, BASF, Kerakoll, Mapei, Sika and Simpson Strong-Tie. These systems are constituted by meshes of different fibres (steel, glass, carbon, basalt and polyparaphenylene benzobisoxazole) and mortars of different composition; nevertheless, they are in general high mechanical performance materials. Recommendations for designing FRP-based strengthening solutions are sufficiently documented in ACI 440 7R-10 [94] and CNR-DT 200 R1/2012 [95] for concrete and masonry elements. These recommendations have also been used for designing TRM strengthening systems for masonry, nevertheless they lack comprehensive instructions, in particular for what concerns application procedures, quality assurance of the materials, durability verification, laboratory and field testing methodologies and the effectiveness of the installation. More recently, the American Concrete Institute published the ACI 549.4R13 [96] to fill this gap.

The use of antiseismic devices to reduce the vulnerability of existing masonry buildings is not frequent, but in the last few years has been gaining attention, especially in the seismic protection of buildings with significant cultural value. Several types of devices have been proposed, among which the most interesting for masonry buildings are base isolators and passive dampers.

Base isolation consists in isolating the structure from the seismic vibrations by introducing an interface with low horizontal stiffness and sufficient capacity to support the vertical loads. This interface modifies the modes of the structure, namely increases their fundamental period and changes their configuration. Furthermore, the increase in horizontal flexibility, introduced by the base isolators, results in a reduction in the accelerations of the structure (by avoiding resonance) and a total displacement increase. Nevertheless, the structure behaves as a rigid body, meaning that the relative displacements of the structure are reduced and thus the associated damage. It should be noted that the total displacements concentrate at the interface of the system, which constitutes a limitation to the use of this solution if they may affect nearby buildings. Base isolation devices can be classified as elastomeric bearings and sliding bearings. Lead–rubber bearings are among the most established devices, whose installation in existing masonry buildings involves (1) construction of reinforced concrete beams at the base of the walls, (2) construction of the base of the devices, (3) installation of the devices, (4) lifting of the structure and transfer of its dead-weight to the devices and (5) construction of a rigid diaphragm at the level of the beams to promote a joint operation of the full system. The lifting–transfer phase is the most critical of the process, as it requires limiting the induced differential settlements to avoid causing damage to the masonry walls and other structural elements. As is implicit, this strengthening solution has an intrusive implementation and especially it can be very costly due to the complexity of the implementation process. Despite this, base isolation has been successfully applied in the strengthening of notable masonry buildings, such as the Salt Lake City and County Building (USA) [97].

The installation of passive dampers presents as main objective the improvement of the structure capacity to dissipate the energy transmitted by an earthquake, which in turn allows to improve the its seismic performance and reduce the induced damage. Dampers can be classified as (1) viscous dampers, (2) viscoelastic dampers, (3) hysteretic metallic dampers and (4) friction dampers. Viscous dampers are constituted by a hollow metallic cylinder filled with a silicon based fluid and a piston that forces the fluid to pass through its small holes at high speed. The movement of the fluid originates high forces that oppose to the relative movement of the damper and dissipates energy through heat due to interparticle friction. The resisting forces are proportional to the piston speed and depend on the viscosity of the fluid. The installation of this type of dampers significantly increases the structural damping, which, in case of an earthquake, allows a significant reduction of the induced deformations. Viscoelastic dampers are normally constituted by metallic sheets intercalated by deformable polymeric sheets. The shear deformation of the polymeric sheets due to relative movements of the damper extremities generates heat and dissipates energy. The dissipated energy is a function of the elastic deformation and deformation speed. In general, the installation of this type of dampers modifies the structural linear parameters for damping and stiffness. Hysteretic metallic dampers dissipate energy by exploiting the plastic deformation capacity of steel, which can be achieved by using low-yielding stress steels or cruciform cross-sections. Finally, friction dampers dissipate energy due to friction generated at the contact surface between two moving solid elements. These dampers are typically constituted by steel plates connected transversally by nuts and bolts fixed with a specific torque, which defines the friction force required to initiate energy dissipation. In general, passive dampers can be installed to strengthen specific structural elements, meaning that they constitute a solution less intrusive and costly than base isolation. The Siena Cathedral (Italy) is an example of a historical monument strengthened with dampers. In this case, two spring viscous dampers were installed to improve the out-of-plane behaviour of the tympanum of the main façade [98].

6. Conclusions

A comprehensive literature review of the main methods and techniques proposed in the last decades to assess and mitigate the seismic vulnerability of existing masonry buildings is presented in this paper. Before making some final remarks, it is worth noting that, despite the authors' efforts to compile a sample of works as relevant and comprehensive as possible, there are certainly several other important works that, due to space constraints or oversight, are not included herein.

From the review given in this paper it is possible to draw some important conclusions. The first regards the selection of the most appropriate seismic vulnerability assessment method to use. As it is mention is Section 2, and easily verifiable from the discussion included in Sections 3 and 4, the selection of the most appropriate method or technique to use must be made based on a proper balance between its simplicity and the accuracy of its results. In this sense, it is very important to stress that vulnerability results can be strongly affected by several sources of uncertainty, which, for this reason, must be acknowledged and addressed. Moreover, and perhaps not less important, in order to be effective, it is fundamental that vulnerability indicators can be easily understood and interpreted, not only by the technical and scientific community, but also by citizens and governmental and civil protection authorities [99]. GIS tools can play a particularly relevant role in this context, making it possible to manage and communicate vulnerability and risk results in a very simple but informative manner. As exemplified in Section 3, such tools can be actually very helpful for the development of strengthening strategies, cost–benefit analyses, civil protection and emergency planning.

The fundamental role of vulnerability assessment methods from the risk mitigation standpoint was also clearly demonstrated though the discussion presented herein. In fact, only on the basis of a comprehensive knowledge of the characteristics of the buildings and of their structural vulnerabilities it is possible to select and implement proficient mitigation strategies and to outline strengthening interventions that can contribute to reduce, in an effective and cost-efficient manner, their seismic vulnerability.

Author Contributions: All authors contributed to every part of the research described in this paper.

Funding: This research was funded by the Portuguese Foundation for Science and Technology (FCT) through the postdoctoral grant SFRH/BPD/122598/2016.

Conflicts of Interest: The authors declare no conflict of interest.

References

1. Whitman, R.V.; Reed, J.W.; Hong, S.-T. Earthquake Damage Probability Matrices. In Proceedings of the Fifth World Conference Earthquakes Engineering, Rome, Italy, 25–29 June 1973.
2. Calvi, G.M.; Pinho, R.; Magenes, G.; Bommer, J.J.; Crowley, H. Development of seismic vulnerability assessment methodologies over the past 30 years. *ISET J. Earthq. Technol.* **2006**, *43*, 75–104.

3. Borzi, B.; Faravelli, M.; Polli, D.A. Central Italy sequence: Simulated damage scenario for the main 2016 shocks. *Bull. Earthq. Eng.* **2018**. [CrossRef]

4. Braga, F.; Dolce, M.; Liberatore, D. A statistical study on damaged buildings and an ensuing review of the MSK-76 scale. In Proceedings of the 7th European Conference on Earthquake Engineering, Athens, Greece, 20–25 September 1982; pp. 431–450.

5. Medvedev, S.W.; Sponheuer, W.; Karnik, V. Seismic intensity scale version MSK 1964. In Proceedings of the First Meeting, Working Group on Seismicity and Seismo-Tectonics, Tbilissi, Georgia, 8–12 June 1965.

6. Ferreira, T.M.; Maio, R.; Costa, A.A.; Vicente, R. Seismic vulnerability assessment of stone masonry façade walls: Calibration using fragility-based results and observed damage. *Soil Dyn. Earthq. Eng.* **2017**, *103*, 21–37. [CrossRef]

7. Di Pasquale, G.; Orsini, G.; Pugliese, A.; Romeo, R. Damage scenario from future earthquakes. In Proceedings of the 11th European Conference on Earthquake Engineering, Paris la Défense, France, 6–11 September 1998.

8. Sieberg, A. Geologie der Erdbeben. *Handb. Geophys.* **1930**, *2*, 552–555.

9. Dolce, M.; Masi, A.; Marino, M.; Vona, M. Earthquake damage scenarios of the building stock of Potenza (Southern Italy) including site effects. *Bull. Earthq. Eng.* **2003**, *1*, 115–140. [CrossRef]

10. Grünthal, G. *European Macroseismic Scale 1998 (EMS-98)*; Cahiers du Centre Européen de Géodynamique et Séismologie: Luxembourg, 1998; Volume 15.

11. Lagomarsino, S.; Giovinazzi, S. Macroseismic and mechanical models for the vulnerability and damage assessment of current buildings. *Bull. Earthq. Eng.* **2006**, *4*, 415–443. [CrossRef]

12. Benedetti, D.; Petrini, V. Sulla vulnerabilita sismica di edifici in muratura: Proposta su un metodo di valutazione. *L'industria delle Costr.* **1984**, *149*, 66–74.

13. GNDT-SSN. *Scheda di Esposizione e Vulnerabilità e di Rilevamento Danni di Primo e Secondo Livello (Murata e Cemento Armato)*; GNDT-SSN: Rome, Italy, 1994. (In Italian)

14. Faccioli, E.; Pessina, V.; Calvi, G.M.; Borzi, B. A study on damage scenarios for residential buildings in Catania city. *J. Seismol.* **1999**, *3*, 327–343. [CrossRef]

15. *ATC-21 Rapid Visual Screening of Buildings for Potential Seismic Hazards*; Federal Emergency Management Agency: Washington, DC, USA, 1988.

16. Bernardini, A. *La Vulnerabilità Degli Edifici: Valutazione a Scala Nazionale Della Vulnerabilità Sismica Degli Edifici Ordinari*; CNR-Gruppo Nazionale per la Difesa dai Terremoti: Roma, Italy, 2000.

17. Petrini, V. Evaluation of risk levels. In *Seismic Hazard and First Evaluation of Risk in Tuscany, CNR—Regione Toscana Technical Report*; Petrini, V., Ed.; Kluwer Academic Publishers: Milano, Italy, 1999. (In Italian)

18. Vicente, R.; D'Ayala, D.; Ferreira, T.M.; Varum, H.; Costa, A.; da Silva, J.A.R.M.; Lagomarsino, S. Seismic Vulnerability and Risk Assessment of Historic Masonry Buildings. In *Structural Rehabilitation of Old Buildings*; Costa, A., Guedes, J.M., Varum, H., Eds.; Springer: Berlin/Heidelberg, Germany, 2014; pp. 307–348, ISBN 978-3-642-39686-1.

19. ATC-13. *Earthquake Damage Estimation Data for California*; Report ATC-13; Applied Technology Council: Redwood City, CA, USA, 1985.

20. Wood, H.O.; Neumann, F. Modified Mercalli intensity scale of 1931. *Bull. Seismol. Soc. Am.* **1931**, *21*, 277–283.

21. Barbat, A.H.; Carreño, M.L.; Pujades, L.G.; Lantada, N.; Cardona, O.D.; Marulanda, M.C. Seismic vulnerability and risk evaluation methods for urban areas. A review with application to a pilot area. *Struct. Infrastruct. Eng.* **2010**, *6*, 17–38. [CrossRef]

22. Eleftheriadou, A.K.; Karabinis, A.I. Development of damage probability matrices based on Greek earthquake damage data. *Earthq. Eng. Eng. Vib.* **2011**, *10*, 129–141. [CrossRef]

23. Fäh, D.; Kind, F.; Lang, K.; Giardini, D. Earthquake scenarios for the city of Basel. *Soil Dyn. Earthq. Eng.* **2001**, *21*, 405–413. [CrossRef]

24. Cardona, O.D.; Yamín, L.E. Seismic Microzonation and Estimation of Earthquake Loss Scenarios: Integrated Risk Mitigation Project of Bogota', Colombia. *Earthq. Spectra* **1997**, *13*, 795–814. [CrossRef]

25. McCormack, T.C.; Rad, F.N. An Earthquake Loss Estimation Methodology for Buildings Based on ATC-13 and ATC-21. *Earthq. Spectra* **1997**, *13*, 605–621. [CrossRef]

26. Maio, R.; Ferreira, T.M.; Vicente, R.; Estêvão, J. Seismic vulnerability assessment of historical urban centres: Case study of the old city centre of Faro, Portugal. *J. Risk Res.* **2016**, *19*, 551–580. [CrossRef]

27. Athmani, A.E.; Ferreira, T.M.; Vicente, R. Seismic risk assessment of the historical urban areas of Annaba city, Algeria. *Int. J. Archit. Herit.* **2018**, *12*. [CrossRef]

28. Lamego, P.; Lourenço, P.B.; Sousa, M.L.; Marques, R. Seismic vulnerability and risk analysis of the old building stock at urban scale: Application to a neighbourhood in Lisbon. *Bull. Earthq. Eng.* **2017**, *15*, 2901–2937. [CrossRef]

29. Maio, R.; Vicente, R.; Formisano, A.; Varum, H. Seismic vulnerability of building aggregates through hybrid and indirect assessment techniques. *Bull. Earthq. Eng.* **2015**, *13*, 2995–3014. [CrossRef]

30. Ferreira, T.M.; Maio, R.; Vicente, R. Analysis of the impact of large scale seismic retrofitting strategies through the application of a vulnerability-based approach on traditional masonry buildings. *Earthq. Eng. Eng. Vib.* **2017**, *16*, 329–348. [CrossRef]

31. Aguado, J.L.P.; Ferreira, T.M.; Lourenço, P.B. The Use of a Large-Scale Seismic Vulnerability Assessment Approach for Masonry Façade Walls as an Effective Tool for Evaluating, Managing and Mitigating Seismic Risk in Historical Centers. *Int. J. Archit. Herit.* **2018**, *12*, 1259–1275. [CrossRef]

32. Lourenço, P.B. Computational Strategies for Masonry Structures. Ph.D. Thesis, Delft University of Technology, Delft, The Netherlands, 1996.

33. Mendes, N.; Lourenço, P.B. Seismic performance of ancient masonry buildings: A sensitivity analysis. In Proceedings of the ECCOMAS Thematic Conference—COMPDYN 2013: 4th International Conference on Computational Methods in Structural Dynamics and Earthquake Engineering, Kos Island, Greece, 12–14 June 2013; pp. 1624–1638.

34. Kurdo, F.A.; Lee, S.C.; Martin, G. Simulating masonry wall behaviour using a simplified micro-model approach. *Eng. Struct.* **2017**, *151*, 349–365. [CrossRef]

35. Andreotti, G.; Graziotti, F.; Magenes, G. Detailed micro-modelling of the direct shear tests of brick masonry specimens: The role of dilatancy. *Eng. Struct.* **2018**, *168*, 929–949. [CrossRef]

36. *DIANA—Finite Element Analysis. DIsplacement Method ANAlyzer*; DIANA FEA BV: Delft, The Netherlands, 2019.

37. ANSYS. *Academic Research Mechanical*; ANSYS, Inc.: Pennsylvania, PA, USA, 2019.

38. *Abaqus Unified FEA*; Dassault Systèmes Simulia Corp.: Providence, RI, USA, 2019.

39. Lourenço, P.B.; Mendes, N.; Ramos, L.F.; Oliveira, D.V. Analysis of masonry structures without box behavior. *Int. J. Archit. Herit.* **2011**, *5*, 369–382. [CrossRef]

40. Mendes, N.; Lourenço, P.B. Seismic Assessment of Masonry "Gaioleiro" Buildings in Lisbon, Portugal. *J. Earthq. Eng.* **2010**, *14*, 80–101. [CrossRef]

41. Betti, M.; Galano, L.; Vignoli, A. Time-History Seismic Analysis of Masonry Buildings: A Comparison between Two Non-Linear Modelling Approaches. *Buildings* **2015**, *5*, 597–621. [CrossRef]

42. Janaraj, T.; Dhanasekar, M. Finite element analysis of the in-plane shear behaviour of masonry panels confined with reinforced grouted cores. *Constr. Build. Mater.* **2014**, *65*, 495–506. [CrossRef]

43. Ferreira, T.M.; Costa, A.A.; Costa, A. Analysis of the Out-Of-Plane Seismic Behavior of Unreinforced Masonry: A Literature Review. *Int. J. Archit. Herit.* **2015**, *9*, 949–972. [CrossRef]

44. Lemos, J.V. Discrete element modeling of masonry structures. *Int. J. Archit. Herit.* **2007**, *1*, 190–213. [CrossRef]

45. Smoljanović, H.; Živaljić, N.; Nikolić, Z. A combined finite-discrete element analysis of dry stone masonry structures. *Eng. Struct.* **2013**, *52*, 89–100. [CrossRef]

46. *UDEC—Distinct Element Modeling of Jointed and Blocky Material in 2D*; ITASCA Consulting Group, Inc.: Minneapolis, MN, USA, 2019.

47. *3DEC—Distinct Element Modeling of Jointed and Blocky Material in 3D*; ITASCA Consulting Group, Inc.: Minneapolis, MN, USA, 2019.

48. *DEMpack*; Center for Numerical Methods in Engineering: Barcelona, Spain, 2019.

49. Mendes, N.; Zanotti, S.; Lemos, J.V. Seismic performance of historical buildings based on Discrete Element Method: An adobe church. *J. Earthq. Eng.* **2018**. [CrossRef]

50. Lemos, J.V.; Campos-Costa, A. Simulation of Shake Table Tests on Out-of-Plane Masonry Buildings. Part (V): Discrete Element Approach. *Int. J. Archit. Herit.* **2017**, *11*, 117–124. [CrossRef]

51. De Felice, G. Out-of-plane seismic capacity of masonry depending on wall section morphology. *Int. J. Archit. Herit.* **2011**, *5*, 466–482. [CrossRef]

52. Tomaževič, M. *The Computer Program POR*; Report ZRMK; Institute for Testing and Research in Materials and Structures: Ljubljana, Slovenia, 1978.

53. *3Muri Software*; S.T.A. DATA: Torino, Italy, 2019.

54. *PRO_SAM, Plugin of the PRO_SAP, ANDIL*; Associazione Nazionale Degli Industriali dei Laterizi: Roma, Italy, 2019.
55. *3DMacro*; Gruppo Sismica s.r.l.: Catania, Italy, 2019.
56. SAP200. *Structural Analysis Program2000*; Computers and Structures, Inc.: Walnut Creek, CA, USA, 2019.
57. Petrovčič, S.; Kilar, V. Seismic failure mode interaction for the equivalent frame modelling of unreinforced masonry structures. *Eng. Struct.* **2013**, *54*, 9–22. [CrossRef]
58. Lagomarsino, S.; Penna, A.; Galasco, A.; Cattari, S. TREMURI program: An equivalent frame model for the nonlinear seismic analysis of masonry buildings. *Eng. Struct.* **2013**, *56*, 1787–1799. [CrossRef]
59. Magenes, G.; Fontana, A.D. Simplified non-linear seismic analysis of masonry buildings. In Proceedings of the 5th International Masonry Conference, London, UK, 12–14 October 1998; British Masonry Society: London, UK, 1998; Volume 1, pp. 190–195.
60. Penna, A.; Lagomarsino, S.; Galasco, A. A nonlinear macroelement model for the seismic analysis of masonry buildings. *Earthq. Eng. Struct. Dyn.* **2014**, *43*, 159–179. [CrossRef]
61. Mendes, N.; Lourenço, P.B. Sensitivity analysis of the seismic performance of existing masonry buildings. *Eng. Struct.* **2014**, *80*, 137–146. [CrossRef]
62. Galaso, A.; Lagomarsino, S.; Penna, A. *TREMURI Program: Seismic Analysis of 3D Masonry Buildings*; Software; University of Genoa: Genoa, Italy, 2002.
63. Orduña, A. Seismic Assessment of Ancient Masonry Structures by Rigid Blocks Limit Analysis. Ph.D. Thesis, University of Minho, Braga, Portugal, 2003.
64. Mendes, N.; Costa, A.A.; Lourenço, P.B.; Bento, R.; Beyer, K.; De Felice, G.; Gams, M.; Griffith, M.; Ingham, J.; Lagomarsino, S.; et al. Methods and approaches for blind test predictions of out-of-plane behavior of masonry walls: A numerical comparative study. *Int. J. Archit. Herit.* **2007**, *11*, 59–71. [CrossRef]
65. Gilbert, M.; Casapulla, C.; Ahmed, H.M. Limit analysis of masonry block structures with non-associative frictional joints using linear programming. *Comput. Struct.* **2006**, *84*, 873–887. [CrossRef]
66. Orduña, A.; Lourenço, P.B. Three-dimensional limit analysis of rigid blocks assemblages. Part I: Torsion failure on frictional interfaces and limit analysis formulation. *Int. J. Solids Struct.* **2005**, *42*, 5140–5160. [CrossRef]
67. Mendes, N. Masonry Macro-block Analysis. In *Encyclopedia of Earthquake Engineering*; Ioannis, M.B., Kougioumtzoglou, A., Patelli, E., Au, I.S., Eds.; Springer: Berlin/Heidelberg, Germany, 2014; ISBN 978-3-642-36197-5.
68. Fajfar, P. A nonlinear analysis method for performance-based seismic design. *Earthq. Spectra* **2000**, *16*, 573–592. [CrossRef]
69. FEMA 440. *Improvement of Nonlinear Static Seismic Analysis Procedures: FEMA 440*; Applied Technology Council (ATC-55 Project) for the Federal Emergency Management Agency (FEMA): Washington, DC, USA, 2005.
70. ASCE. *Seismic Evaluation and Retrofit of Existing Buildings: ASCE/SEI 41-13*; Virginia American Society of Civil Engineers: Reston, VA, USA, 2013.
71. Azizi-Bondarabadi, H.; Mendes, N.; Lourenço, P.B. Higher Mode Effects in Pushover Analysis of Irregular Masonry Buildings. *J. Earthq. Eng.* **2019**, in press. [CrossRef]
72. Chácara, C.; Mendes, N.; Lourenço, P.B. Simulation of Shake Table Tests on Out-of-Plane Masonry Buildings. Part (IV): Macro and Micro FEM Based Approaches. *Int. J. Archit. Herit.* **2017**, *11*, 103–116. [CrossRef]
73. Housner, G.W. The behavior of inverted pendulum structures during earthquakes. *Bull. Seismol. Soc. Am.* **1963**, *53*, 403–417.
74. Sorrentino, L.; D'Ayala, D.; De Felice, G.; Griffith, M.C.; Lagomarsino, S.; Magenes, G. Review of Out-of-Plane Seismic Assessment Techniques Applied to Existing Masonry Buildings. *Int. J. Archit. Herit.* **2017**, *11*, 2–21. [CrossRef]
75. Ortega, J.; Vasconcelos, G.; Rodrigues, H.; Correia, M.; Lourenço, P.B. Traditional earthquake resistant techniques for vernacular architecture and local seismic cultures: A literature review. *J. Cult. Herit.* **2017**, *27*, 181–196. [CrossRef]
76. Moreira, S.; Ramos, L.F.; Oliveira, D.V.; Lourenço, P.B. Design parameters for seismically retrofitted masonry-to-timber connections: Injection anchors. *Int. J. Archit. Herit.* **2016**, *10*, 217–234. [CrossRef]
77. Vintzileou, E. Effect of timber ties on the behavior of historic masonry. *J. Struct. Eng.* **2008**, *134*, 961–972. [CrossRef]

78. Michiels, T.L. Seismic retrofitting techniques for historic adobe buildings. *Int. J. Archit. Herit.* **2015**, *9*, 1059–1068. [CrossRef]

79. Moreira, S.; Ramos, L.F.; Oliveira, D.V.; Lourenço, P.B. Experimental behavior of masonry wall-to-timber elements connections strengthened with injection anchors. *Eng. Struct.* **2014**, *81*, 98–109. [CrossRef]

80. Paganoni, S.; D'Ayala, D. Testing and design procedure for corner connections of masonry heritage buildings strengthened by metallic grouted anchors. *Eng. Struct.* **2014**, *70*, 278–293. [CrossRef]

81. Senaldi, I.; Magenes, G.; Penna, A.; Galasco, A.; Rota, M. The effect of stiffened floor and roof diaphragms on the experimental seismic response of a full-scale unreinforced stone masonry building. *J. Earthq. Eng.* **2014**, *18*, 407–443. [CrossRef]

82. Binda, L. The difficult choice of materials used for the repair of brick and stone masonry walls. In Proceedings of the 1st International Conference on Restoration of Heritage Masonry Structures, Cairo, Egypt, 24–27 April 2006.

83. Oliveira, D.V.; Silva, R.A.; Garbin, E.; Lourenço, P.B. Strengthening of three-leaf stone masonry walls: An experimental research. *Mater. Struct.* **2012**, *45*, 1259–1276. [CrossRef]

84. Hurd, J. Observing and applying ancient repair techniques to pisé and adobe in seismic regions of Central Asia and Trans-Himalaya. In Proceedings of the Getty Seismic Adobe Project 2006 Colloquium, Getty Conservation Institute, Los Angeles, CA, USA, 11–13 April 2006; pp. 101–108.

85. Valluzzi, M.R.; Binda, L.; Modena, C. Mechanical behaviour of historic masonry structures strengthened by bed joints structural repointing. *Constr. Build. Mater.* **2005**, *19*, 63–73. [CrossRef]

86. Karakostas, C.; Lekidis, V.; Makarios, T.; Salonikios, T.; Sous, I.; Demosthenous, M. Seismic response of structures and infrastructure facilities during the Lefkada, Greece earthquake of 14/8/2003. *Eng. Struct.* **2005**, *27*, 213–227. [CrossRef]

87. Toumbakari, E. Lime-Pozzolan-Cement Grouts and their Structural Effects on Composite Masonry Walls. Ph.D. Thesis, Catholic University of Leuven, Leuven, Belgium, 2002.

88. Mazzon, N.; Valluzzi, M.R.; Aoki, T.; Garbin, E.; De Canio, G.; Ranieri, N.; Modena, C. Shaking table tests on two multi-leaf stone masonry buildings. In Proceedings of the 11th Canadian Masonry Symposium, Toronto, ON, Canada, 31 May–3 June 2009.

89. Illampas, R.; Silva, R.A.; Charmpis, D.C.; Lourenço, P.B.; Ioannou, I. Validation of the repair effectiveness of clay-based grout injections by lateral load testing of an adobe model building. *Constr. Build. Mater.* **2017**, *153*, 174–184. [CrossRef]

90. Papanicolaou, C.G.; Triantafillou, T.C.; Papathanasiou, M.; Kyriakos, K. Textile Reinforced Mortar (TRM) versus FRP as Strengthening Material of URM Walls: Out-of-plane Cyclic Loading. *Mater. Struct.* **2008**, *41*, 143–157. [CrossRef]

91. Valluzzi, M.R.; Modena, C.; De Felice, G. Current Practice and Open Issues in Strengthening Historical Buildings with Composites. *Mater. Struct.* **2014**, *47*, 1971–1985. [CrossRef]

92. De Felice, G.; De Santis, S.; Garmendia, L.; Ghiassi, B.; Larrinaga, P.; Lourenço, P.B.; Oliveira, D.V.; Paolacci, F.; Papanicolaou, C.G. Mortar-based Systems for Externally Bonded Strengthening of Masonry. *Mater. Struct.* **2014**, *47*, 2021–2037. [CrossRef]

93. Ascione, L.; De Felice, G.; De Santis, S.A. Qualification Method for Externally Bonded Fiber Reinforced Cementitious Matrix (FRCM) Strengthening Systems. *Compos. Part B Eng.* **2015**, *78*, 497–506. [CrossRef]

94. *ACI 440.7R-10: Guide for the Design and Construction of Externally Bonded FRP Systems for Strengthening Unreinforced Masonry Structures*; American Concrete Institute: Farmington Hills, MI, USA, 2010.

95. *CNR-DT200 R1/2012: Guide for the Design and Construction of Externally Bonded FRP Systems for Strengthening Existing Structures*; Italian Council of Research (CNR): Rome, Italy, 2012; pp. 1–167.

96. *ACI 549.4R13: Guide to Design and Construction of Externally Bonded Fabric-Reinforced Cementitious Matrix (FRCM) Systems for Repair and Strengthening Concrete and Masonry Structures*; American Concrete Institute (ACI): Farmington Hills, MI, USA, 2013; ISBN 9780870318528.

97. Bailey, J.S.; Allen, E.W. Seismic isolation retrofitting: Salt Lake City and county building. *APT Bull. J. Preserv. Technol.* **1988**, *20*, 33–44. [CrossRef]

98. Castellano, M.G.; Tosti, G.; Bolletti, G.P.; Tosti, M. *Fluid Spring Dampers for Seismic Protection of the Cathedral of Siena*; Protection of Historical Buildings—PROHITECH: Rome, Italy, 2009; pp. 669–674.

99. Maio, R.; Ferreira, T.M.; Vicente, R. A critical discussion on the earthquake risk mitigation of urban cultural heritage assets. *Int. J. Disaster Risk Reduct.* **2018**, *27*, 239–247. [CrossRef]

buildings

Article

Action Protocols for Seismic Evaluation of Structures and Damage Restoration of Residential Buildings in Andalusia (Spain): "IT-Sismo" APP

Emilio J. Mascort-Albea [1,*], Jacinto Canivell [2], Antonio Jaramillo-Morilla [1], Rocío Romero-Hernández [1], Jonathan Ruiz-Jaramillo [3] and Cristina Soriano-Cuesta [1]

[1] Department of Building Structures and Soil Engineering, Higher Technical School of Architecture, Universidad de Sevilla, 41012 Sevilla, Spain; jarami@us.es (A.J.-M.); rociorome@us.es (R.R.-H.); csoriano@us.es (C.S.-C.)

[2] Department of Architectural Construction II, Higher Technical School of Building Engineering, Universidad de Sevilla, 41012 Sevilla, Spain; jacanivell@us.es

[3] Department of Art and Architecture, Architectural Construction´s Chair, Higher Technical School of Architecture, Universidad de Málaga, 29071 Málaga, Spain; jonaruizjara@uma.es

* Correspondence: emascort@us.es; Tel.: +34- 954-556-662

Received: 22 March 2019; Accepted: 25 April 2019; Published: 29 April 2019

Abstract: The seismotectonic conditions of the Iberian Peninsula trigger the occurrence of earthquakes with an occasional periodicity, but with intensities greater than VI on the European macroseismic scale (EMS). For this reason, local action protocols are required in order to efficiently organise the technical inspections that must be carried out on a massive scale after events such as the earthquakes experienced in the Spanish cities of Lorca (2011) and Melilla (2016). This paper proposes the development of a set of documents for the evaluation and diagnosis of the state of existing buildings and infrastructure regarding seismic activity in Andalusia. With special attention paid to residential typology, approximations have been carried out to the normative context, to general comparatives, to particular analyses of a case studies selection, and to complementary approaches. The results have led to the establishment of two specific protocols. Firstly, the short-term guideline enables the classification of damage and risk levels, and the determination of what immediate interventions should be carried out through the generation of a preliminary on-site report. This activity can be performed by architects and engineers with the help of a mobile-device application (APP IT-Sismo Andalucía). Additionally, a long-term protocol provides calculation procedures and constructive solutions for the improvement of the seismic behaviour of affected buildings. Specially designed tests demonstrate the validity of the protocols and illustrate the need for information and communication technologies (ICT) tools in the evaluation of architectonic technical aspects.

Keywords: mobile-device applications; automatic protocols; damage assessment; in situ structural diagnosis; seismic restoration

1. Introduction

The seismotectonic conditions of the Iberian Peninsula provoke the intermittent occurrence of earthquakes with intensity values higher than IV on the European macroseismic scale (EMS) [1]. Within this geographic framework, the autonomous community of Andalusia presents especially high risk values in the areas linked to the mountain ranges of the Baetic system (Sierra Bética). This issue can be checked analysing the earthquake recurrence data in this area [2,3], which is one of the most populated regions in Spain with more than eight million inhabitants and an area close to 90,000 km^2.

In this respect, the unfortunate consequences of this type of seismic event have been experienced in recent years in populations very close to the Andalusian territory. In 2011, the city of Lorca, located in the border region of Murcia, suffered the effects of an earthquake of magnitude 5.1 and intensity VII (EMS), while in 2016, the city of Melilla, located on the African continent, suffered an earthquake of magnitude 6.3 that was felt with intensity III (EMS). In the specific case of Lorca, there were nine deaths, 310 injuries, and more than a thousand damaged buildings. Moreover, it should be borne in mind that the damage resulting from construction deficiencies, mainly landslides, were greater than the damage caused by structural failures in the affected buildings [4,5]. The earthquake that occurred in Melilla left no fatalities, but the constructive damage experienced was nevertheless very high [6–8]. In both cases, residential buildings constituted an architectural type especially affected by seismic events, which, in Andalusia, pose a threat to an estimated four million homes [9].

In the Portuguese territories closest to the Andalusian region, there is also a clear scenario of vulnerability to earthquakes. In this context, various studies have been developed for the evaluation of the seismic vulnerability of buildings located in urban centres, such as Faro, Lisbon, Coimbra, and Oporto. Such studies have shown that certain areas with a relatively low hazard level can present a high seismic risk due to a lack of maintenance of their buildings and to recent restoration interventions that increase the possibility of suffering damage from a seismic phenomenon [10–12].

Based on the aforementioned circumstances, the implementation of strategies to raise awareness both of the public and of public administrations regarding seismic events should be considered a priority task. This work is crucial not only for suitable decision-making, but also for raising awareness of the scope of repercussions caused by a relevant seismic event [13]. In this respect, in 2009, the Emergency Plan for Seismic Risk in Andalusia [14] was published at a regional level. Additionally, the seismic hazard map of the province of Malaga [15] at the provincial and local level, and the Municipal Action Plans for the seismic risk of Granada [16] and Benalmádena [17] have been published. These documents mainly focus on risk-reduction strategies, on the detection of the most highly vulnerable areas, and on planning for seismic damage scenarios through the adoption or proposal of specific procedures. These documents include a series of recommendations for people on how to proceed in the event of an earthquake and, in certain cases, they provide brief indications in order to perform a preliminary assessment of the damage and risks produced, and information regarding the safety measures to be employed.

These documents clearly show the growing concern of technicians and local authorities regarding the ever-present lack of seismic protection at both regional and local levels. Therefore, specific action protocols still need to be established for those technicians and citizens who can contribute towards the evaluation of the scope and the real risks of damage caused after the occurrence of a seismic event. This type of document is not only aimed at ensuring the safety of users, but also at detecting high-risk situations when carrying out technical inspections, in which professionals are exposed to aftershocks and to the existence of constructive elements that are in a state of unstable equilibrium. In this regard, at the regional level and in the period between 2006/2007, the project titled SISMOSAN, an assessment of the seismic hazard and risk in Andalusia [18] was developed. Subsequently, the authors of this article carried out a research project funded through a public call for bids, under the title "Seismic standard: Preliminary analysis and restoration in the face of damage to existing buildings and infrastructures". On completion of said project, a phase of testing and adjustment of the results was carried out, based on the digital tools generated within the framework of the aforementioned work.

Through this project, the design of a set of action guidelines has been analysed in greater depth. In a complementary and coherent way, these guidelines are aimed at expediting not only the mechanisms of technical and citizen mobilization in anticipation of future seismic events of relevant intensity that may occur in the Andalusian area, but also any measures for immediate action. Although these procedures could be applied in a general way to any building type, the tools developed are used for the evaluation of residential buildings, which are highly representative at a quantitative level. Therefore, specific aspects related to the use, typology and construction have been considered, within such a broad theme as that of housing.

In this way, this proposal develops a protocol for the evaluation and diagnosis after an earthquake of the Andalusian residential stock. All the planned tasks have been carried out through the application of procedures validated by the main international institutions in this field. At the same time, as a measure of methodological innovation, an interactive validation instrument is provided in the form of an application for mobile devices (APP), which, based on the use of information and communication technologies (ICT), favours decision-making by architects and engineers not specialized in seismic issues. In this way, the developed APP enables the post-catastrophe evaluation sheets to be filled out interactively and provides a preliminary automated diagnosis based on the information supplied by the technicians who supervise the properties affected by a seismic event.

This question is of major interest since, from the requirements demanded in each normative update, degrees of seismic behaviour can be inferred from certain groups of residential types in a general way, and therefore common technical and constructive aspects can be established in the analysis of the buildings.

The aim of this paper is to present a security needs assessment protocol to evaluate residential building after an earthquake, using a two-step procedure depending on the complexity of the situation, the availability of resources, and the size of the affected area. In a first step, a general survey would be carried out to assess the main damages, based on the technician's observations, and if it is needed; in a second step, a detailed analysis can be performed, which is usually more time-consuming. This document is especially focused in the development of the first step of the analysis.

2. Methodology and Resources Employed

By means of a structure of contents that enables the creation of a set of documents capable of responding to the complexity of the proposed objectives, the following lines of study and analysis have been developed in parallel (Figure 1): normative approximation and international guidelines, principal methods of evaluation, context conditions, general characteristics of Andalusian residential construction, and representativeness of the case studies. To this end, a multidisciplinary work team has been formed, mainly composed of architects and civil engineers. Furthermore, the most representative contents have been structured in an application for mobile devices, which is destined to function as an interactive guide for the use of the newly established protocols. For this purpose, a team of computer engineers has created a computer application based on the structure proposed by the work team, which will be discussed in Section 3.1.1. All this has led to the development of research, which, through a holistic and generalist approach, has addressed numerous issues by establishing a broad view of the problem at hand, and has contributed towards the progressive improvement of the current situation through the use of various approaches and complementary perspectives.

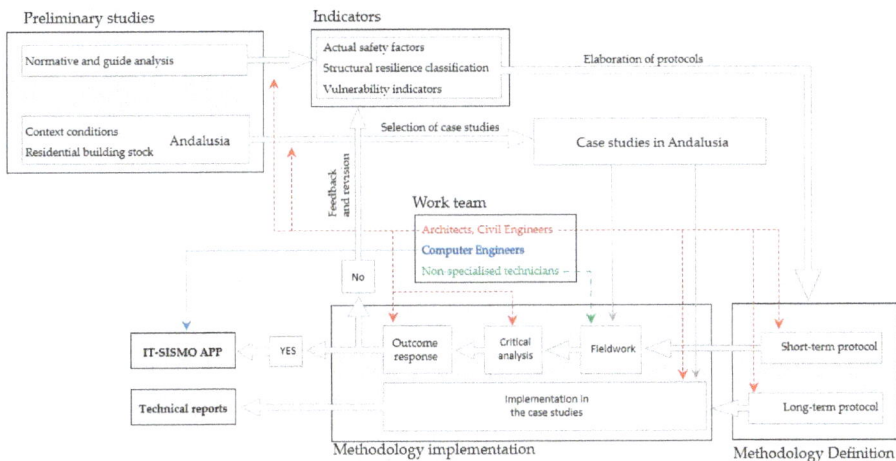

Figure 1. General outline relative to the development of the methodology of the present research. Source: Prepared by the authors.

2.1. Normative Approach and International Guidelines

In order to indicate those gaps and areas of opportunity that have been covered in the development of this work, a detailed analysis is carried out on the national regulations and seismic guidelines published in Spain. In this respect, the first Spanish normative text that refers to the seismic issue is "Chapter 7. Seismic Actions, of the MV-101 Standard on Actions in Construction" [19], published in 1962 with the objective of considering this type of event in the context of constructive design. Since this initial contribution, the influence of seismic effects has been taken into account in the "Instructions for the project, construction, and exploitation of large dams of the Ministry of Public Works" (IGP-1967) [20], and in the first state regulation of specific character published in 1968 under the title "Seismic-resistant Standard PGS-1/Part A". The aforementioned document never reached completion, and in 1974 it was replaced by the Seismic-resistant Standard PDS-1 [21]. In that same year the Permanent Commission of Seismic-resistant Standards was founded for the purpose of revising the seismic standard every five years. However, no subsequent normative update was made until twenty years later, with the publication of the "Seismic-resistant Construction Standard: General Part and Construction" (NCSE-1994) [22]. This document has been updated with the NCSE-2002 [23] for the general field of construction and with the "Seismic-resistant Construction Standard: Bridges" (NCSP-2007) [24], which is principally oriented towards the field of civil engineering. Both documents remain in force across Spain today, thereby illustrating that the five-year regulatory review period initially proposed is not currently being met.

The improvements provided in these updates have been found to provide criteria and specifications aimed at decisively influencing the quality of the residential buildings that have been built under the new regulations [25,26]. However, taking into account the normative panorama as indicated, a set of deficiencies can be detected, related to the actions in restoration work and to the evaluation of damage in catastrophic situations, that present a wide margin for documentary improvement. At the normative level, in Europe, the different Eurocodes and in the case study of Eurocode 8: Project of seismic-resistant structures [27], and more specifically, the various National Annexes particular to each of the countries, provide a series of common methods for the calculation of the mechanical strength of the elements that play a structural role in construction work.

In order to propose procedures that suppose an advance in the degree of detail of the norms and codes currently published in Spain, an approximation has been made to the panorama of seismic-resistance guidelines and protocols of post-catastrophe action that are published at an international level. Special attention has been paid to those countries and regions that have a greater documentary production in this subject. Most of these documents offer a series of general recommendations related to prompt actions. These recommendations are aimed at expediting the decision-making of the various institutional managers. In this respect, work by the Federal Emergency Management Agency (FEMA) in the United States [28,29] is of special interest, as is that by the New Zealand Society for Earthquake Engineering (NZSEE) [30,31], whose influence transcends publications in other countries with major seismic events, such as those in Japan [32]. The publication of guidelines in South America and Central America is also very common, with a high level of detail offered by the seismic codes in countries such as Chile, Colombia, and Mexico [33,34]. A large part of these documents is inspired by the requirements of the ACI-318, [35] as published by the "American Concrete Institute," the 2019 revision of which is about to be published. Its latest publication was in 2015 and included a version in Spanish and Chinese. In Europe, the development and importance acquired by these types of guidelines is of note [36], in countries such as Portugal, Italy, Greece, and in Turkey in the regions of Mediterranean influence.

The criteria and considerations collected in the study of the aforementioned documents highlight the importance that should be awarded to preventive studies that achieve the best possible development of "post-catastrophe" actions. Furthermore, their content has proven to be especially useful for the development of the short-term action protocol, primarily in the design and configuration of those parameters that form part of the readily prepared inspection records, as indicated in Section 3.1 of this document.

2.2. Main Methods of Evaluation

Based on the set of standards and guidelines consulted in the document analysis process illustrated in the preceding section, the main evaluation methods taken into account in the development of this work are those related to the configuration of parameters destined for the diagnosis of the state of conservation of a property following a seismic phenomenon. These methods are considered without undermining the importance and validity of those indicators related to the current level of risk that buildings endure in the face of future earthquakes. The main methods of analysis considered in the development of the project are specified below (Figure 2):

- Vulnerability indices (pre-catastrophe): Among the main indicators used in the risk-prevention study, that of the definition of the vulnerability index (Iv) stands out [10,11,37–39]. In general terms, the index is evaluated by weighting a series of quantitative and qualitative parameters that describe the current state of a building and its structural system. Its collection is based on the indications established by the GNDT II level [40], whose usefulness is largely associated with evaluation strategies prior to the occurrence of catastrophic phenomena, in order to prevent and minimize seismic damage. This method is based on the Risk-EU methodology [41]. This procedure quantitatively develops the vulnerability assessment proposal in accordance with the structural typology included in the European Macroseismic Scale [42]. In contrast, the first method of evaluation enables greater precision in the prediction of possible damage, by establishing a classification of the buildings and zones in accordance with the particular or general levels of vulnerability obtained.

- Diagnoses of immediate action (post-catastrophe): The principal methods and guidelines consulted opt for the use of simple and direct diagnoses that enable a clear determination of what the immediate action should be after a seismic event. This decision is associated with the extent of the seismic effect on the building, which can be determined through its general state of conservation, and through the state presented by its specific structural and constructive elements. The most usual casuistry proposes the establishment of three degrees of action: "apparently safe use", "restricted use", and "dangerous use". Through this determination, a first filter is achieved related to the functionality of the buildings after the catastrophe.

- Safety coefficients calculated from real values (pre- and post-catastrophe): Other studies focus on numerical analyses that assess the safety factor of the buildings after an earthquake. These procedures are also known as direct techniques [11] and, given the high degree of knowledge required of the property to be evaluated, they tend to be used for the diagnosis of specific buildings, especially those classified as historical heritage [42,43], buildings with specific typologies with a high degree of typological and constructive homogeneity [44], and those that house general services whose operation is essential in case of catastrophe, such as hospitals [45–48]. This type of analysis usually requires a more detailed knowledge of the particular properties of the building, which are not always available directly, and often involve a major effort in prior characterization thereof. In turn, the determination of the safety coefficients is conditioned by the regulations in force at each geographical location. However, the application of this type of method provides not only a complete and highly detailed view of the degree of particular resilience that each analysed building can present, but also the margin of safety that it may attain. In this respect, its contributions for the study of vernacular typologies in Portugal [44] and for the development of evaluations carried out in Italy [49] deserve mention.
- Classifications of structural strength assessment (pre- and post-catastrophe): This methodological approach, derived from the results obtained in the evaluation discussed above, constitutes an additional step that includes the process of calculating safety factors in specific buildings. In this respect, it is interesting to use classifications that refer to the level of structural safety (or quality) of a property against possible seismic phenomena, through the use of codes that allow buildings to be labelled in a similar way to the system employed for the energy rating of buildings. Similar classification procedures are currently used to estimate the degree of conservation of buildings in historical centres, thereby establishing priority for their intervention [50].

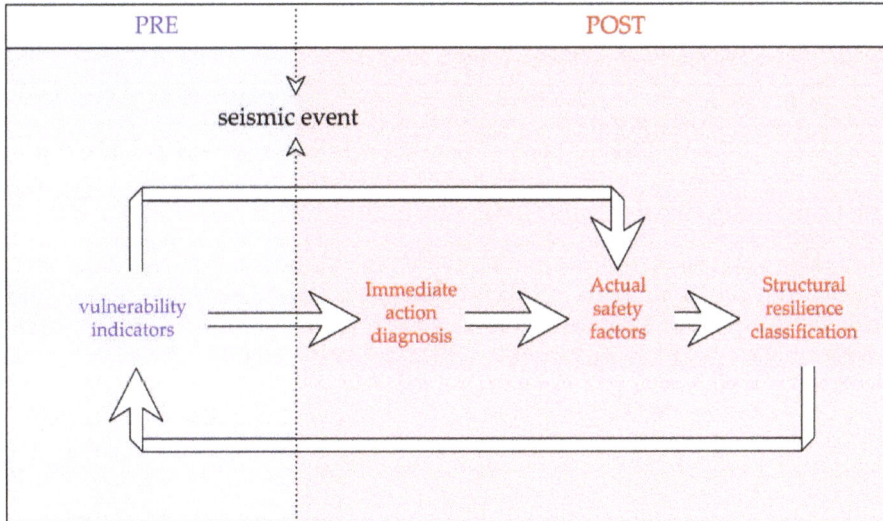

Figure 2. Diagram of the main analytical methods used in the research. Source: Prepared by the authors.

In order to carry out the evaluation of a building after an earthquake, it is essential, as a starting point, to have previous information about its safety conditions. The different procedures described above, allow a global analysis of the building, developing a structural characterization with a multiple approach. On the one hand, the risk-UE methodology provides the seismic vulnerability information in an easy and direct way, according to its constructive, morphological and typological characteristics. This information must be improved highlighting those points of the structure that a priori present greater vulnerability, that should be the object of special attention after the earthquake. This method, based on real values (dimensions of beams, material properties or reinforcements), provides an objective factor to assess safety: the higher the safety factor, the lower the risk of damage to the element analysed during an earthquake. Additionally, the procedures for immediate diagnoses (post-catastrophe) and classifications of structural strength assessment (pre- and post-catastrophe) have provided information on agile and effective inspection procedures, in order to estimate the structural response of a building. Thus, methodology procedures and a general and specific look over the building are combined to assess the survey and the evaluation after the earthquake. Through this approach, it can be observed how the application of these methods must be carried out through an iterative process that acquires validity in the episodes both before and after a seismic catastrophe. Each episode provides information of interest to achieve the various objectives, included in a common framework whose purpose is the optimization of resources and of management measures when faced with this type of catastrophe.

The vulnerability approach explained above, along with the local soil properties has been applied to the Andalusian region through a micro-scale analysis. The different methodologies provide indicators that can be plotted through micro-zoning studies, useful and necessary for a better control of the actual situation on the architectural scale. For this purpose, the use of geographic information systems (GIS) can be considered especially valid [51]. These systems, as collaborative instruments towards the management of this information through the open use of data, facilitate not only delocalised work between the technicians and professionals involved, but also the opening of communication channels that contribute towards the diffusion of the information generated.

2.3. Context Conditions

The effect of the amplification of seismic waves in terms of the type of terrain establishes the local seismic hazard; this is known in seismic engineering as the site effect. Although several authors have conducted research to determine the amplification factors caused by this effect [52], in order to analyse the influence of seismic and geotechnical conditions of the terrain in constructions located in Andalusia, the main values and criteria specified in the NSCE-02 have been used; these regulations are currently in force at national level. This document requires the detailed characterization of the soil to a depth of 30 metres for the calculation of the seismic coefficient (C). This information, in spite of its obligatory nature, is seldom collected in the realization of geotechnical tests in building work. Indeed, several studies carried out in Lorca after the 2011 earthquake [53] have shown the importance of the study of the site at a geological level for the determination of the level of risk in the event of an earthquake, whereby a major part of the damage is derived from the lax consideration of its importance.

In order to define the geotechnical conditions of the Andalusian subsoil, additional material has been retrieved from the geological and geotechnical maps published at national level by institutions such as the Spanish Geo-Mining Institute (IGME). In this sense, regional studies related to the preparation of a geotechnical map of Andalusia have also been employed using GIS technology (Figure 3). A series of 1:400,000 scale maps have been produced by the project team, providing general information. On the one hand, thirty-eight basic soil units have been established for Andalusia, and on the other, the special conditions associated with karstic and expansive soils and landslides have been identified. All this information has been mapped in detail in a further set of cartographies that, using the 1:50,000 scale, provide in-depth information on the geotechnical characteristics of each Andalusian province.

Finally, this territorial approach has been accomplished with the help of institutional publications that address the study of risks in Andalusia [54]. Through these guidelines it is possible to corroborate the existence of zones that present a special seismic risk and that are geographically associated with the Baetic mountain system.

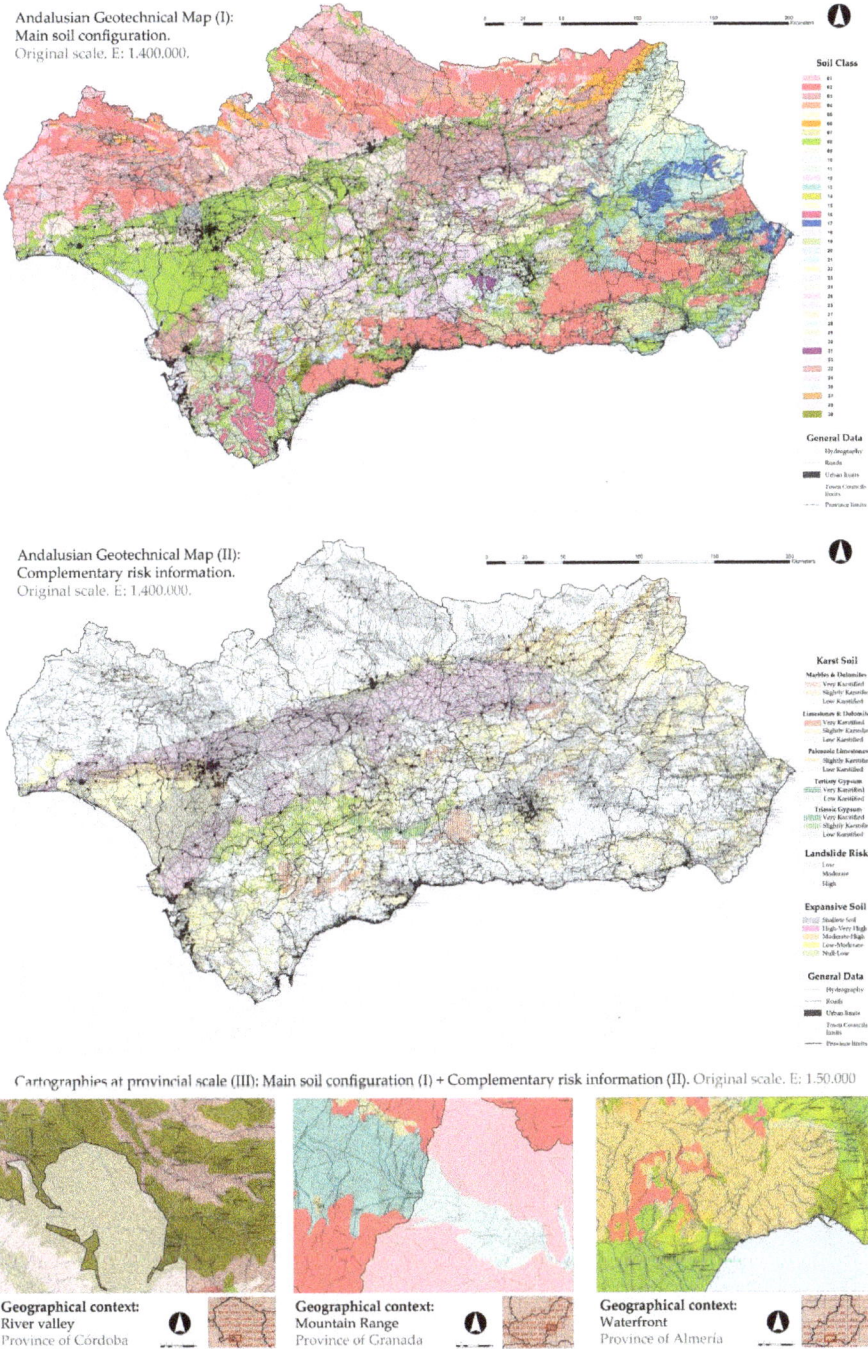

Figure 3. Geotechnical Map of Andalusia. General cartography and examples of detailed maps. Source: Prepared by the authors.

The analysis of this documentation has brought to light the existence of large documentary and informational gaps in terms of data for the calculation of seismic conditions at urban and architectural level. Although local studies have been carried out in provincial capitals such as Malaga [55], the micro-zoning work in the Andalusian territory cannot be considered to be general; there is therefore a deficit in the quantity and quality of this information.

Therefore, complementary to the assessment of the seismic vulnerability of residential buildings, and within the framework of the present investigation, specific maps of seismic micro-zoning have been drawn, based on the collected data, that take into account the evaluation of possible liquefaction phenomena in spaces representative of historical centres, as is the case of Seville [56].

2.4. General Characteristics of Andalusian Residential Buildings

In relation to the specific topic of this work, a general study of the typological and constructive conditions of the housing in Andalusia has been made. The main analyses have been developed according to the structural and constructive systems used in different periods, complied from several Spanish regulations. In this respect, it is necessary to highlight the prevalence of the use, since the 20th century, of structural and constructive systems largely based on the production of bricks, in periods prior to the establishment of the first seismic regulations. Steel and especially reinforced concrete then became the most common structural typologies, compared to other materials such as rammed earth and wood. It is precisely through the most common architectural applications of these systems that a graphic catalogue has been constructed which enables the determination of the main damage that can be experienced by elements built following an event of seismic origin.

Moreover, it should be borne in mind that the large volume of residential stock was built before the second half of the 20th century. All those buildings that have yet to undergo a comprehensive review lack the protection of some type of regulation, code, or set of guidelines that pay specific attention to their possible structural and constructive behaviour in the event of an earthquake. Hence, it can be considered that this group of buildings is characterized by a recognizable heterogeneity and should be studied with a special degree of attention in future research related to the topic addressed herein.

2.5. Representativeness of the Case Studies

In order to analyse the characterization tasks in greater depth, the seismic-resistant conditions of a representative selection of case studies have been evaluated in detail, through the preparation of reports of seismic vulnerability relative to specific buildings. The properties have been selected based on their locations and are associated with geographical situations that may constitute a significant sample of the behaviour of certain residential typologies. Thus, from among the case studies analysed, residential buildings located in each of the eight provinces of the Andalusian region have been included, with priority granted to the study of collective housing blocks, especially those related to the public offer and built on the basis of revisable regulations. Furthermore, it should be borne in mind that preliminary information has been provided by the Agency for Housing and Restoration of Andalusia (AVRA), thereby contributing to the creation of the reports set forth in this subsection.

The case study reports have been written according to a general structure that enables a common working guide to be established for the inspection and seismic restoration of buildings. In this way, the analysed aspects in the different technical documents include the main issues that have subsequently been developed in the design of the short- and long-term protocols. As a result of this work, this methodology has been applied to different particular conditions that allow its general validity to be verified. Below, Table 1 provides a detailed description of the structure of the technical reports.

Table 1. General organization of the case study reports. Source: Prepared by the authors.

Chapter	Main Contents of Each Chapter
(1) Introduction	background; previous documentation; location; site-data; chronology; general risks conditions
(2) Seismic Vulnerability Factors	seismic regulations; seismic information; soils data; geometry of the built-up complex
(3) Constructive Description	foundation; structure; enclosures and partitions; roofs
(4) Building Survey	damage survey (cracks, material loosening); non-destructive tests (ground-penetrating radar, accelerometers, deformation leveling)
(5) Intervention Proposals	reinforcements; underpinning; bracings

Consequently, the drafted documents provide a constructive and structural diagnosis of the case studies analysed, through the application of the main evaluation methods laid out in Section 2.2. In this way, the calculation of the real safety coefficients of the selected buildings has been carried out (Figure 4), by using finite element models (FEM) with the software SAP2000. The evaluation of the sections of their main structural elements has also been drawn by using the computer application of the Concrete Catalogue EHE 2008.

The results obtained using this software have allowed to establish patterns of structural behaviour for the typology studied. Thus, those specific points of the structure with greater vulnerability have been revealed. The use of linear static analysis has been considered suitable since the proposed methodology does not intend to be an exhaustive structural verification. More complex procedures such as nonlinear static analysis, linear dynamics analysis, nonlinear dynamic analysis are time consuming and their level of precision is not necessary in order to develop a preliminary assessment about the safety of a building after an earthquake. Additionally, when the situation has required it, non-destructive tests have been carried out to obtain additional information to characterize the structural properties of the buildings through the use of ground-penetrating radar, environmental measurements with accelerometers, or vertical deformation levelling [57].

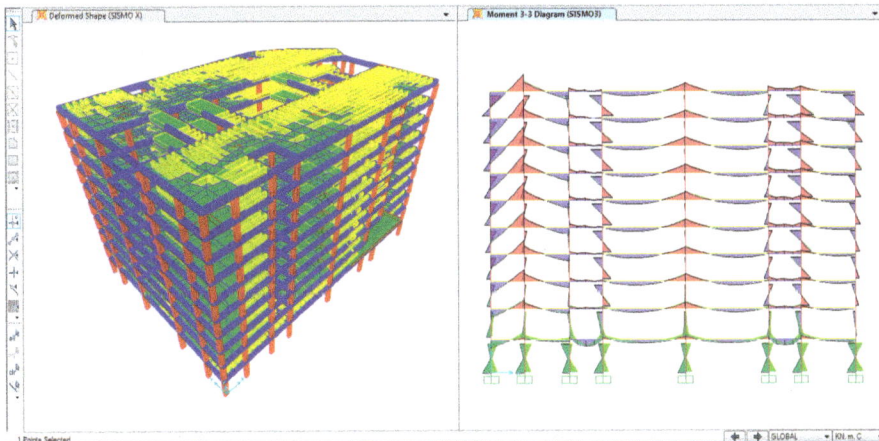

Figure 4. Digital model for the calculation by finite elements of officially protected housing built in 2003, located in C/ América 2, Cádiz (Spain). Source: Prepared by the authors.

Through this work, reference values have been obtained, suitable for similar cases. In turn, a structure for detailed analysis is provided that can be perfectly applicable for the seismic evaluation of numerous residential properties of similar characteristics.

3. "Post-Catastrophe" Seismic Evaluation Protocol for Residential Buildings in Andalusia

The structure of the information is organized into two main categories (Figure 5), which enable the consideration of the inspection activities and damage classification criteria that must be taken into account (short-term protocol). It also enables the illustration of the calculation procedures and constructive solutions that must be applied for the development of subsequent seismic restoration projects (long-term protocol).

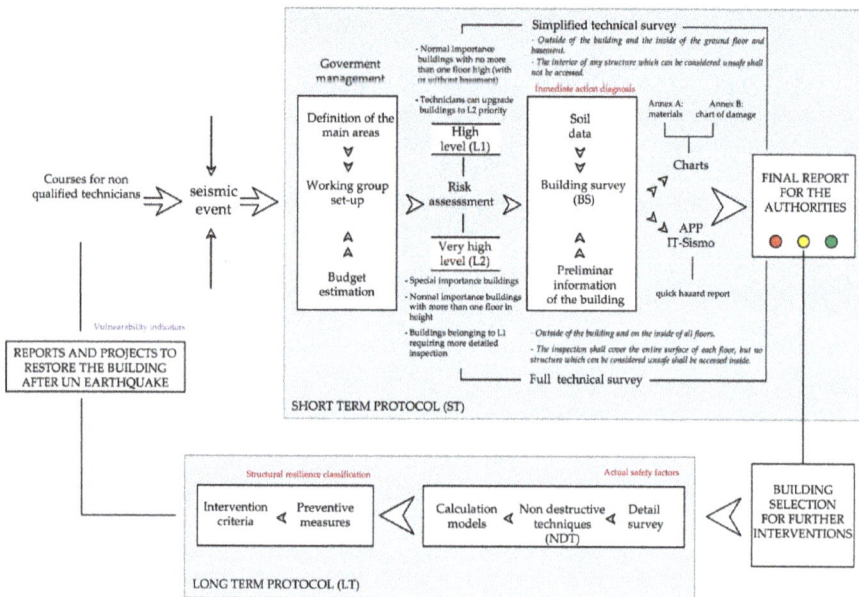

Figure 5. Process of evaluation and relation with short- and long-term protocols. Source: Prepared by the authors.

In this way, it is observed how all the preventive studies that have been developed throughout the project are materialized in a set of procedures and instruments aimed at the implementation of "post-catastrophe" actions.

At a general level, the use of the application for data collection in the analysis of the state of each of the buildings following a seismic event, would allow the management of all the reports in real time. Thus, the information related to the evaluated buildings would be included in a common platform that, based on GIS procedures, would allow the creation of urban maps in which to visualize the areas with the highest incidence of damage. This would facilitate the organization of the actions of the emergency teams following the earthquake, especially regarding short-term protocol.

3.1. Short-Term Action Protocol: Prompt Post-Earthquake Assessment Guidelines

The Protocol of short-term action is intended for the immediate inspection of residential buildings. This procedure includes a prompt action plan where the first guidelines for control, inspection, analysis, and evaluation of the state of the buildings affected by a major category earthquake are established. This document allows the stability of the inspected homes to be analysed and their degree of functional hazard to be determined depending on the damage detected. Once a short period of time has elapsed since the action of an earthquake, guidelines are employed to evaluate the state of material conservation of the built-up area and that of its structural and constructive elements. In order to ascertain the degree of safety of buildings following the disaster, a set of inspection checklists have been prepared based on the prescriptions provided by the main international guidelines [28,30,36].

Applying basic recommendations, the designed charts aid the classification of the damages in the building and enable the determination of its safety after the earthquake (Figure 6). These resources are designed to be used in quick assessments of damaged buildings, giving information about their degree of deterioration and avoiding more human lives lost. Therefore, the terms "Simplified technical survey" and "Full technical survey" are related to the resources used and the information provided in each type of inspection. This type of inspection is related to the "High" (L1) and "Very High" (L2) priority levels, which are established depending on the characteristics of the buildings checked and the need to repeat certain inspections, according to the obtained results. Through the consideration of the points included in this protocol, at a general level, the most sensitive points collected should be analysed in immediate inspections. These points include the following: Instability of the building due to the soil (settling, soil liquefaction, landslides, etc.); collision between buildings due to insufficient joints for seismic movement; breakage of short pillars in semi-basements; breakage of pillars on the ground floor due to changes in stiffness; loose façade elements and structural joints failure.

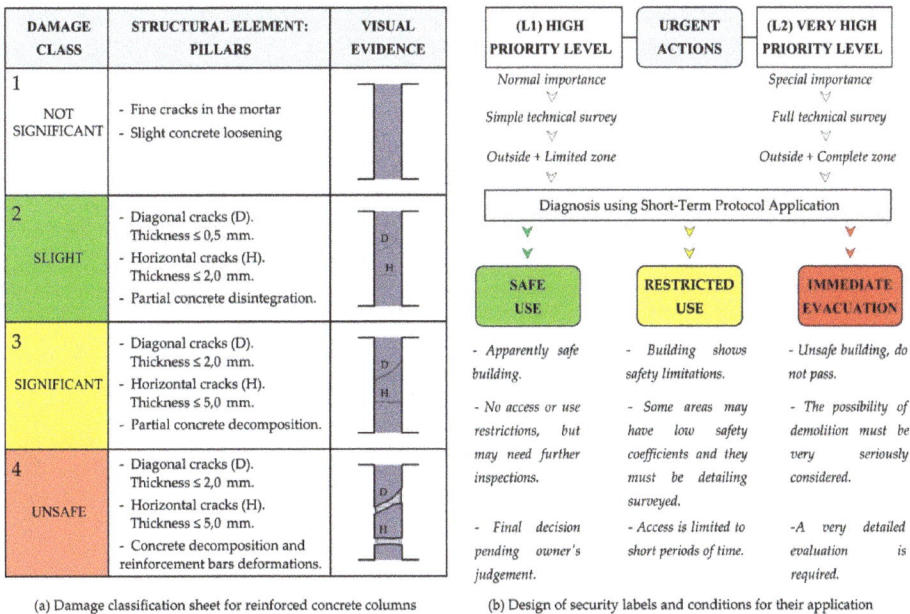

DAMAGE CLASS	STRUCTURAL ELEMENT: PILLARS	VISUAL EVIDENCE
1 NOT SIGNIFICANT	- Fine cracks in the mortar - Slight concrete loosening	
2 SLIGHT	- Diagonal cracks (D). Thickness ≤ 0,5 mm. - Horizontal cracks (H). Thickness ≤ 2,0 mm. - Partial concrete disintegration.	
3 SIGNIFICANT	- Diagonal cracks (D). Thickness ≤ 2,0 mm. - Horizontal cracks (H). Thickness ≤ 5,0 mm. - Partial concrete decomposition.	
4 UNSAFE	- Diagonal cracks (D). Thickness ≤ 2,0 mm. - Horizontal cracks (H). Thickness ≤ 5,0 mm. - Concrete decomposition and reinforcement bars deformations.	

(L1) HIGH PRIORITY LEVEL	URGENT ACTIONS	(L2) VERY HIGH PRIORITY LEVEL
Normal importance ⋎		*Special importance* ⋎
Simple technical survey ⋎		*Full technical survey* ⋎
Outside + Limited zone ⋎		*Outside + Complete zone* ⋎

Diagnosis using Short-Term Protocol Application

SAFE USE	RESTRICTED USE	IMMEDIATE EVACUATION
- *Apparently safe building.* - *No access or use restrictions, but may need further inspections.* - *Final decision pending owner's judgement.*	- *Building shows safety limitations.* - *Some areas may have low safety coefficients and they must be detailing surveyed.* - *Access is limited to short periods of time.*	- *Unsafe building, do not pass.* - *The possibility of demolition must be very seriously considered.* - *A very detailed evaluation is required.*

(a) Damage classification sheet for reinforced concrete columns   (b) Design of security labels and conditions for their application

Figure 6. Damage classification sheets and measures to be taken depending on the inspection results. Source: Prepared by the authors.

Once all the above points have been considered, the technician responsible for the inspection is guided by the information included in the document regarding the results of the diagnosis in relation to the actual safety level of the residential building inspected after the occurrence of the seismic event. In this respect, three general levels are provided [31]: Safe (green label: the building has minor damage, there is no access restriction); Restricted Use (yellow label: affected areas are detected, passage is prohibited except for extraordinary circumstances); Immediate Evacuation (red label: damage that can affect the stability of the building, passage is forbidden).This document is designed to guide the technicians responsible for carrying out technical inspections in their decisions as to which principal factors and parameters should be considered during said inspections: work teams, geometry and location of the building, constructive elements and technical requirements of the regulations in force. This set of guidelines is proposed as an instrument for professionals in the field of architecture and engineering, who, without being specialized in seismic events, can perform collaborative work, essential under this type of circumstance, with greater knowledge.

In order to facilitate the use of the contents of the protocol in the short term, the possibility of designing an interactive tool was proposed for the automatic completion of the inspection forms. Based on this objective, the IT-Sismo Andalucía APP was created, which can be considered as one of the most innovative contributions of the present research.

3.1.1. Application for IT-Sismo Andalucía Mobile Devices

As indicated above, the contents of the short-term protocol have been integrated into an application for mobile devices (cell phones and tablets) in the Spanish language. This application provides interactive tools for the development of the aforementioned diagnosis and the possibility of producing drawings and taking photographs that could be included in the final evaluation report. This document is generated automatically, following the completion of the fields of information that appear in the sequence offered by the digital tool.

Among the various options of functionality that have been considered, it was decided, in consensus with the technicians of the supervising institutions, that the application could work without an internet connection. In this way, the APP becomes operative through a simple process of prior download and subsequent installation on the inspecting technician's device. This is carried out in anticipation of the possible problems of connectivity and access to data that may occur in the event of a catastrophic situation. With this approach, it is intended to eliminate a possible dependence that could be experienced by the technicians in charge of the various inspections.

The first contact of the user with the application occurs via the presentation screen, which provides an introduction regarding the characteristics of the project, as well as a set of recommendations aimed at facilitating its use. To this end, the contents of the protocols explained above are provided, as well as a manual for the tool itself. Once the users have become familiarised with the aforementioned contents, they will be able to use the interactive protocol to good advantage (Figure 7).

Figure 7. Several screenshots of the ICT tool developed for the assessment of existing buildings and infrastructures (IT-Sismo). Source: Prepared by the authors.

The tool enables the creation of diagnostic and evaluation reports, and also the editing, deletion, and downloading of all the preliminary versions that the user deems appropriate. Once the results achieved are verified as being those expected by the technicians, the generated contents can be validated through their signature. This option implies a blocking of the generated report, which remains stored in the user's personal file, and can be sent to the competent institutions. All the reports are structured into three large blocks, which must be completed consecutively:

- General information: Initially, the technician completes the data in a first phase by providing general information related to the work team, the building, and the soil upon which its foundations lie. On the screen related to the work team, information relating to the type of inspection carried out can be included as can the contact details of all members who make up the inspection squad which must be coordinated by a technician responsible for verifying the final document. The section related to the property data requires the definition of its location, its chronology, its volumetric characteristics, its materiality, and the nature of its typology. Through the last screen, related to the soil-structure interaction, the tool automatically contributes the values related to the seismic coefficients of the soil (*ab/g* and K) from the site indicated, and allows the introduction of the values of the magnitude and intensity experienced during the earthquake. These parameters are used by the APP to classify the type of damage and are usually related to the occurrence of each kind of seismic event. Likewise, the definition of the main constructive and structural characteristics that influence the interaction with the soil is considered, as are the values of the fundamental period of the building that could be measured with sensors integrated in mobile devices.

- Inspection: The information provided in this first block allows the tool to supply a series of recommendations for the development of the second phase of the work, related to the detailed inspection of the architectural elements of the building. From this phase, a set of fields are activated that relate to the different materials introduced in the previous phase. In this way, independent interactive forms are established that work as guidelines for the quantification of the percentages of damage experienced with respect to all the constructive and structural elements of the building that have been fabricated in concrete, steel, stonework, and/or wood. Concerning these material categories, a second differentiation is established between resistant elements, auxiliary elements, foundation failures, and general movements. In addition, and due to their importance in the evaluation process, the resistant elements are again categorized through the following concepts: beams, pillars, walls, joints, floor slabs, and stairs. In those categories of most common damage, the tool provides graphic diagrams and images that allow the user to visually identify the type of damage referenced in the field to be filled. Additionally, the application allows photographs to be taken and drawings to be made through the mobile device itself, which can be stored and included as documentation attached to the report generated.
- Damage Assessment: In order to quantify the damages to the building and its risks, the different elements are classified in groups based on their typology, from A to D, and each of them is given a label depending on the severity of the damage (classes from 1 to 4). Combining the severity of damages and its spread percentage, the building will be labelled in a different group as the following chart shows (Table 2).

Table 2. Standards for the assessment of particular elements. Source: Prepared by the authors.

Group	Description	Damage Class	Damage Spread (%)	Group Label
A.1	Main structural elements: beams, pillars, joints, load-bearing walls, etc.	1–2	25–50	Green
		1–2	75–100	Yellow
		3	25–50	Yellow
		3	75–100	Red
		4	25–100	Red
A.2	Horizontal or inclined structural elements: slabs, floors, stairs, roofs, etc.	1–2	25–50	Green
		1–2	75–100	Yellow
		3	25–50	Yellow
		3	75–100	Red
		4	25–100	Red
B.1	Non-resistant elements: masonry walls, fillings, partition walls, enclosures, etc.	1–2	25–100	Green
		3	25–50	Green
		3	75–100	Yellow
		4	25–50	Yellow
		4	75–100	Red

Table 2. *Cont.*

Group	Description	Damage Class	Damage Spread (%)	Group Label
B.2	Auxiliary elements: parapets, chimneys, roofing elements, tiles, coverings, glassware, lighting, antennas, etc.	1–2	25–50	Green
		1–2	75–100	Yellow
		3	25–50	
		3	75–100	Red
		4	25–100	
C	Building verticality	1–2	Not applicable	Green
		3	Not applicable	Yellow
		4	Not applicable	Red
D	Soil and foundations	1	Not applicable	Green
		2	Not applicable	Yellow
		3–4	Not applicable	Red

By means of empirical rules and qualitative methods that have been checked and validated by experts, a specific label is assigned to the building depending on the combination of the amount of damage and its spread. The following chart determines the overall assessment of the affected building according to the degree of damage established in each group of elements (Table 3).

Table 3. General building assessment and label criteria. Source: Prepared by the authors.

Groups	Group Label				Proposed Group Combination	Building Label
A.1 or A.2 or B.1	Red				1	Immediate evacuation
A.1 or A.2 or B.1 and B.2	Yellow and Green				2	Restricted use
A.1 or A.2 or B.1 and B.2	Yellow and Red				3	Immediate evacuation
A.1 and A.2 and B.1 and B.2 and C or D	Green and Yellow				4	Restricted use
A.1 and A.2 and B.1 and B.2 and C or D	Green and Red				5	Immediate evacuation
A.1 and A.2 and B.1 and B.2 and C or D	Yellow and Yellow	or	Red		6	Immediate evacuation
A.1 and A.2 and B.1 and B.2 and C and D	Green and Yellow or Green	or	Red and		7	Safe use
A.1 and A.2 and B.1 and B.2 and C and D	Green Yellow Yellow	or or	Red Red		8	Restricted use
A.1 and A.2 and B.1 and B.2 and C and D	Green				9	Safe use

The aforementioned procedure has been implemented in the IT-Sismo App, so that once each indicator is typed in, the software generates the label. The evaluation of the collected data through the app allows the architect or engineer to correctly label the building. In any case, the technician will be able to edit the results obtained and include comments that help understand the document, prior to its final validation.

3.2. Long-Term Intervention Protocol: Guide to Seismic Restoration

This document is proposed as a guide for the creation of technical reports that require prolonged development over time. This document presents the work and actions that technicians must take into account in order to carry out a more detailed analysis, regarding the restoration and/or demolition of the properties that require this type of revision. In this way, the scope and objectives of this document are related to the contribution of criteria for the evaluation of seismic behaviour in existing buildings, the description of possible corrective measures to be used in a restoration process, and the orientation in the development of tasks related to the analysis of the structural types, the recalculation of the affected elements, and the sizing proposals of the new parts to be incorporated in reinforcement and/or refitting operations.

The long-term intervention protocol establishes the systems that check and verify existing structures. These systems are necessary for the analysis of the behaviour of buildings if, where applicable, they have to face an earthquake of high intensity. This document is aimed towards orienting the calculation of the current structural safety coefficient of the constructions before dynamic action, towards evaluating their degree of conservation, and towards proposing new safety indices, especially of either those buildings whose typological and constructive characteristics are representative of most of the domain of the Andalusian region, or of constructions of a patrimonial nature. In this case, the conditions of buildings of a historical nature are tackled.

Here, specific calculation and modelling criteria are taken into account for the structural evaluation of the housing, as are structural and constructive solutions for the reinforcement and repair of damaged elements. One example includes the establishment of new safety coefficients and highly specific test systems, such as those based on the determination of the natural frequency of vibration.

The protocol will be accompanied by a series of technical recommendations, and by possible actions of reparation, restoration, and reinforcement, aimed at improving the capacity of the buildings to be resistant to future earthquakes, whereby as little as possible of the previous configuration is altered, and increasing its safety in the event of an earthquake [32].

4. Discussion

With the arrival of the 21st century, a gradual awareness has arisen regarding historical seismic risk, especially in regions where the incidence of earthquakes of high magnitude and intensity is rare. This is the case of Andalusia, in the south of Spain, where historical records reveal the occurrence of earthquakes with an intensity of IX or higher; such as the earthquakes of 1504, 1755, and 1884 [58,59]. Due to its long recurrence interval, there has traditionally been a lack of awareness of this risk. This level of awareness has increased in recent years, whereby certain municipalities have developed action plans in case disaster strikes, although this preparation has yet to be generalized. These plans not only should include a set of action protocols, but they must also incorporate specific procedures regarding the inspection of damaged buildings and the way to proceed in the management of the information generated.

After a seismic event of an intensity greater than VI, it is necessary to carry out an inspection of the properties and affected areas in order to detect their degree of safety. Given the workload, it is imperative to have a specific tool that enables a quicker assessment and also the operational administration of the amount of information generated, so the most effective decisions for the management of the possible catastrophe can be adopted in real time. In this sense, those strategies designed for the development of such a complex documentary set, established as a "post-catastrophe" protocol on a regional scale, have been combined in a proposal of a methodological nature. This is achieved through the development of an electronic application installed on a mobile device, that serves as a guide for the inspection, and which also enables the establishment of a regulated procedure for data collection.

This tool can be used by the technicians in charge of carrying out the inspections and allows all the information to be downloaded onto a common platform which aids the management and decision-making process by the authorities.

The specific information included in the short-term protocol comes from the analysis of the general constructive conditions of residential buildings in Andalusia, and from examples of damage taken from the studies collected in this work. In order to expedite an inspection process that must be carried out urgently, an application for mobile devices has been developed for the automation of diagnostics. Fortunately, no seismic event has yet led this work to be put directly into practice. However, testing has been carried out through a verification of the coherence and logical value of the results. To this end, eight case studies, namely eight residential buildings developed by the public administration, have been introduced in the App and evaluated. These results were compared with a prior manual evaluation of researchers and technicians for the same buildings and by means of improvement cycles, the procedures were adapted until the outcome of the App matched the manual evaluation. The proposed methodology has been tested in numerous case studies proving its effectiveness during a post-earthquake evaluation. To do this, based on the selected group of residential buildings, multiple damage scenarios have been designed and applied to buildings with different structural and constructive characteristics. All the results obtained have been contrasted by the authors of this work and researchers of the project to verify the suitability and logic of the results obtained. This has allowed to optimize the use of the APP.

Furthermore, the diagnosis reports applied to the selected case studies have enabled the calculation conditions to be verified for the creation of the protocol in the long term; a set of guidelines has been designed systematizing those aspects that a seismic restoration proposal for residential buildings in Andalusia should include.

Additionally, the tool has been shared in teaching sessions with students and non-specialist technicians, demonstrating the usability and viability of the digital version of the protocol through the IT Sismo APP. This illustrates the usefulness of ICT tools in the evaluation of technical aspects that require a high degree of specialization.

5. Conclusions

The proposed methodology has been developed to assess residential buildings considering a two-step analysis. The first step of the evaluation concerns the preliminary work that has been proved to be easy to implement for any residential building, and most importantly, it can be used by non-experts with few resources, so that in case of seismic event it is a more efficient way to cover the evaluation of a greater number of affected cases.

It should be pointed out that the creation of reports in digital format, as an outcome of the IT-Sismo APP, facilitates the inspection and speeds up the institutional management of the post-catastrophe files that are generated during the process. This procedure also allows the management of the information obtained, thereby increasing the level of efficiency by reducing the time allocated to data collection and the preparation of a comprehensive diagnosis.

As this innovative proposal is the first of this kind in Andalusia, it should be considered as an initial approach; hence the methodology can be improved in certain ways. For example, other structural types or construction buildings materials need to be evaluated, since within the building stock in Andalusia we can find many of them dated before the 20th century, and therefore a new perspective on the traditional constructive systems should be considered. As this proposal is a result of a public funded project, the authors suggest that this tool should be implemented within the workflow of the administrations in charge of the maintenance of the building stock, either by means of the off-line tool or a future on-line one.

Buildings **2019**, *9*, 104

Finally, this proposal should also serve as a reminder that society must remain alert (and prepared) for the occurrence of phenomena, that in Spain (and more specifically, in Andalusia) are rare, but could also be catastrophic.

Author Contributions: All the authors of this publication, collectively, have contributed to the development of the following tasks: conceptualization; methodology; software; validation; formal analysis; investigation; data curation; writing (original draft preparation); writing (review and editing); supervision; project administration; funding acquisition.

Funding: The results achieved in the research are the result of the development of the project "Norma sísmica. Análisis previo y rehabilitación de edificios e infraestructuras existentes", (Seismic standard. Previous analysis and restoration of existing buildings and infrastructures), financed by the *Programa Operativo FEDER de Andalucía 2007–2013*. Additionally, the writing of this document has been financed with resources of V Plan Propio de Investigación of the University of Seville.

Acknowledgments: The authors of this work are grateful for the technical supervision of the project carried out by Manuel Páez Antúnez and Dragana Kovandzic Mlacenovic, who belong to different agencies of the Ministry of Development and Housing of the Junta de Andalucía. Additionally, it is necessary to highlight the collaboration of Marta Mora Santisteban as a member of the project team. Finally, the participation of Andrea Cimino Arriaga and Antonio Pérez Sánchez in the technological development of the computer application is especially appreciated.

Conflicts of Interest: The authors declare that there are no conflicts of interest in this work.

References

1. Grünthal, G. (Ed.) *European Macroseismic Scale 1998 (EMS-98)*; Cahiers du Centre Européen de Géodynamique et de Séismologie 15, Centre Européen de Géodynamique et de Séismologie: Luxembourg, 1998.
2. Morales-Esteban, A.; Martínez-Álvarez, F.; Reyes, J. Earthquake prediction in seismogenic areas of the Iberian Peninsula based on computational intelligence. *Tectonophysics* **2013**, *593*, 121–134. [CrossRef]
3. Martínez-Álvarez, F.; Reyes, J.; Morales-Esteban, A.; Rubio-Escudero, C. Determining the best set of seismicity indicators to predict earthquakes. Two case studies: Chile and the Iberian Peninsula. *Knowl.-Based Syst.* **2013**, *50*, 198–210. [CrossRef]
4. Navarro, M.; García-Jerez, A.; Alcalá, F.J.; Vidal, F.; Enomoto, T. Local site effect microzonation of Lorca town (SE Spain). *Bull. Earthq. Eng.* **2014**, *12*, 1933–1959. [CrossRef]
5. Alguacil, G.; Vidal, F.; Navarro, M.; García-Jerez, A.; Pérez-Muelas, J. Characterization of earthquake shaking severity in the town of Lorca during the May 11, 2011 event. *Bull. Earthq. Eng.* **2014**, *12*, 1889–1908. [CrossRef]
6. Casado, C.L.; Garrido, J.; Delgado, J.; Peláez, J.A.; Henares, J. HVSR estimation of site effects in Melilla (Spain) and the damage pattern from the 01/25/2016 Mw 6.3 Alborán Sea earthquake. *Nat. Hazards* **2018**, *93*, 153–167. [CrossRef]
7. Buforn, E.; Pro, C.; de Galdeano, C.S.; Cantavella, J.V.; Cesca, S.; Caldeira, B.; Udías, A.; Mattesini, M. The 2016 south Alboran earthquake (Mw = 6.4): A reactivation of the Ibero-Maghrebian region? *Tectonophysics* **2017**, *712–713*, 704–715. [CrossRef]
8. Cortés, A.E.N. *How Is Perceived the Environment by the Newspapers? A Brief Approach*; ARKEOS: Lima, Perú, 2017; pp. 31–38.
9. Ministerio de Fomento. *Gobierno de España*; Estimación del parque de viviendas: Madrid, España, 2017.
10. Maio, R.; Ferreira, T.M.; Vicente, R.; Estêvão, J. Seismic vulnerability assessment of historical urban centres: Case study of the old city centre of Faro, Portugal. *J. Risk Res.* **2016**, *19*, 551–580. [CrossRef]
11. Vicente, R.; Parodi, S.; Lagomarsino, S.; Varum, H.; Silva, J.A.R.M.; Vicente, R.; Varum, H.; Parodi, S.; Langomarsino, S.; Silva, J.A.R.M. Seismic vulnerability and risk assessment: Case study of the historic city centre of Coimbra, Portugal. *Bull Earthq. Eng.* **2011**, *9*, 1067–1096. [CrossRef]
12. Ferreira, T.M.; Vicente, R.; da Silva, J.A.R.M.; Varum, H.; Costa, A. Seismic vulnerability assessment of historical urban centres: Case study of the old city centre in Seixal, Portugal. *Bull. Earthq. Eng.* **2013**, *11*, 1753–1773. [CrossRef]
13. Vicente, R.; Ferreira, T.M.; Maio, R.; Koch, H. Awareness, Perception and Communication of Earthquake Risk in Portugal: Public Survey. *Procedia Econ. Financ.* **2014**, *18*, 271–278. [CrossRef]
14. Junta de Andalucia. *Plan de Emergencias ante el Riesgo Sísmico en Andalucía*; Consejería de Presidencia, Administración Pública e Interior: Sevilla, Spain, 2009.

15. Bernal, J.A.C.; Arribas, R.G.; Alcón, F.G. Mapa de peligrosidad sísmica de la provincia de Málaga. In *Planificación y medidas de actuación ante un seísmo o terremoto*, 1st ed.; Diputación de Málaga: Málaga, Spain, 2017.

16. *Plan de Actuación municipal ante el riesgo sísmico de Granada (PLAMSIGra)*, 1st ed.; Ayuntamiento de Granada: Granada, Spain, 2016.

17. *Plan de actuación local ante riesgo sísmico en Benalmádena; Ayuntamiento de Benalmádena*; Dirección de Gestión de Emergencias: Granada, Spain, 2012; p. 115.

18. Gaspar-Escribano, J.M.; Navarro, M.; Benito, B.; García-Jerez, A.; Vidal, F. From regional- to local-scale seismic hazard assessment: Examples from Southern Spain. *Bull. Earthq. Eng.* **2010**, *8*, 1547–1567. [CrossRef]

19. Ministerio de la Vivienda. *Norma MV-101/1962*; Acciones en la Edificación. Decreto 195/1963; Ministerio de la Vivienda, Servicio Central de Publicaciones: Madrid, Spain, 1963.

20. Ministerio de Obras Públicas. Instrucción para proyecto, construcción y explotación de grandes presas (IPG-1967) (1967). Spain. Available online: https://www.boe.es/buscar/doc.php?id=BOE-A-1967-17302 (accessed on 10 April 2019).

21. Ministerio de Planificación del Desarrollo. Norma Sismorresistente P.D.S-1, Pub. L. No. *BOE-A-1974-186 (1974). Spain.* Available online: https://www.boe.es/boe/dias/1974/11/21/pdfs/A23585-23601.pdf (accessed on 10 April 2019).

22. Comisión Permanente de Normas Sismorresistentes. *Norma de Construcción Sismorresistente: Parte General y Edificación (NCSE-94)*; Boletín Oficial Del Estado (BOE), 35898-35967; Dirección General del Instituto Geográfico Nacional: Madrid, Spain, 1994.

23. Comisión Permanente de Normas Sismorresistentes. (2002). Norma de Construcción Sismorresistente: Parte General y Edificación (NCSE-02). Serie normativas. España: Ministerio de Fomento. Gobierno de España. Available online: https://www.fomento.gob.es/MFOM.CP.Web/handlers/pdfhandler.ashx?idpub=BN0222 (accessed on 10 April 2019).

24. Comisión Permanente de Normas Sismorresistentes. *Norma de Construcción Sismorresistente: Puentes (NCSP-07)*; Boletín Oficial Del Estado (BOE): Madrid, Spain, 2007; Volume 132, pp. 24044–24133.

25. Martínez, R.B. Enfoque y avances conceptuales de la nueva Norma Española de Construcción Sismorresistente NCSE-94. *Inf. La Construcción* **1997**, *48*, 39–45. [CrossRef]

26. Barbat, H.; Oller, S.; Vielma, J.C. *Cálculo y diseño sismorresistente de edificios*; Aplicación de la norma NCSE-02, Centro Internacional de Métodos Numéricos en Ingeniería: Barcelona, España, 2005.

27. JRC. *EN 1998: Design of Structures for Earthquake Resistance*; Eurocode 8, Pub. L. No. EN 1998-1; European Committee for Standardization: Brussels, Belgium, 2004.

28. Federal Emergency Management Agency (FEMA). *FEMA-310. Handbook for the Seismic Evaluation of Buildings*; FEMA: Washintong, DC, USA, 1998.

29. Federal Emergency Management Agency (FEMA). *FEMA-273. NEHRP Guidelines for the Seismic Rehabilitation of Buildings*; FEMA: Washintong, DC, USA, 1997.

30. New Zealand Society for Earthquake Engineering (NZSEE). *Field Guide: Rapid Post Disaster. Building Usability Assessment—Earthquakes*; New Ministry of Business, Innovation and Employment: Wellingtong, New Zealand, 2015.

31. New Zealand Society for Earthquake Engineering (NZSEE). *Building Safety Evaluation During a State of Emergency*; Guidelines for Territorial Authorities; Department of Building and Housing: Wellingtong, New Zealand, 2009.

32. Nakano, Y.; Maeda, M.; Kuramoto, H.; Murakami, M. Guideline for post-earthquake damage evaluation and rehabilitation of buildings in Japan. In Proceedings of the 13th Conference on Earthquake Engineering, Vancouver, BC, Canada, 1–6 August 2004; pp. 1–15. [CrossRef]

33. Barrientos, M. Terremotos, arquitectura y la Norma Chilena NCh433. Of.72. Cálculo Antisísmico de Edificios (1959-1972). In *Anales de la Arquitectura. 2017-2018*; Ediciones ARQ: Santiago de Chile, Chile, 2018; pp. 10–19.

34. García, L.E. Desarrollo de la Normativa Sismo Resistente Colombiana en los 30 años desde su primera expedición. *Rev. Ing.* **2015**, *71*. [CrossRef]

35. ACI. *318-14: Building Code Requirements for Structural Concrete and Commentary*; ACI-318; American Concrete Institute: Farmington Hills, MI, USA, 2014.

36. Anagnostopoulos, S.A.; Moretti, M.; Panoutsopoulou, M.; Panagiotopoulou, D.; Thoma, T. *Post Earthquake Damage and Usability Assessment of Buildings: Further Development and Applications*; Final report; European Commission: Brussels, Belgium, 2004.

37. Shakya, M.; Varum, H.; Vicente, R.; Costa, A. Seismic vulnerability assessment methodology for slender masonry structures. *Int. J. Archit. Herit.* **2018**, *12*, 1297–1326. [CrossRef]

38. Shakya, M. GNDT II level approach for seismic vulnerability assessment of unreinforced masonry (URM) building stock. *J. Sci. Eng.* **2015**, *3*, 21–27. [CrossRef]

39. Azizi-Bondarabadi, H.; Mendes, N.; Lourenco, P.; Sadeghi, N. A seismic vulnerability index method for masonry schools in the province of Yazd, Iran. In Proceedings of the 9th International Masonry Conference, Guimarães, Portugal, 7–9 July 2014.

40. GNDT–SSN. *Scheda di esposizione e vulnerabilità e di rilevamento danni di primo livello e secondo livello (muratura e cemento armato)*; INGV. Gruppo Nazionale per la Difesa dai Terremoti: Roma, Italy, 1994.

41. Milutinovic, Z.V.; Trendafiloski, G.S. *WP4: Vulnerability of Current Buildings*, 1st ed.; European Commission: Brussels, Belgium, 2003.

42. Augusti, G.; Ciampoli, M.; Giovenale, P. Seismic vulnerability of monumental buildings. *Struct. Saf.* **2001**, *23*, 253–274. [CrossRef]

43. Barbieri, G.; Biolzi, L.; Bocciarelli, M.; Fregonese, L.; Frigeri, A. Assessing the seismic vulnerability of a historical building. *Eng. Struct.* **2013**, 523–535. [CrossRef]

44. Barros, R.; Rodrigues, H.; Varum, H.; Costa, A.; Correia, M. Seismic Analysis of a Portuguese Vernacular Building. *J. Archit. Eng.* **2018**, *24*, 05017010. [CrossRef]

45. Cardona, O.D. Vulnerabilidad sísmica de hospitales. In *Fundamentos para ingenieros y arquitectos*, 848992533X ed.; Centro Internacional de Métodos Numéricos en Ingeniería: Barcelona, Spain, 1999.

46. Vona, M.; Anelli, A.; Mastroberti, M.; Murgante, B.; Santa cruz, S. Prioritization Strategies to reduce the Seismic Risk of the Public and Strategic Buildings. *Disaster Adv.* **2017**, *10*, 1–5.

47. Grant, D.N.; Bommer, J.J.; Pinho, R.; Calvi, G.M.; Goretti, A.; Meroni, F. A Prioritization Scheme for Seismic Intervention in School Buildings in Italy. *Earthq. Spectra* **2007**, *23*, 291–314. [CrossRef]

48. Grimaz, S.; Slejko, D. Seismic hazard for critical facilities. *Bollettino di Geofisica Teorica ed Applicata* **2014**, *55*, 3–16. [CrossRef]

49. Bernardini, A.; D'Ayala, D.; Modena, C.; Speranza, E.; Valluzzi, M. Vulnerability assessment of the historical masonry building typologies of Vittorio Veneto (NE Italy). *Bollettino di Geofisica Teorica ed Applicata* **2008**, *49*, 463–484.

50. Piñero, I.; San-José, J.T.; Rodríguez, P.; Losáñez, M.M. Multi-criteria decision-making for grading the rehabilitation of heritage sites. Application in the historic center of La Habana. *J. Cult. Herit.* **2017**, *26*, 144–152. [CrossRef]

51. Zarzosa, M.N.L. Evaluación del riesgo sísmico mediante métodos avanzados y técnicas GIS. Ph.D. Thesis, Universidad Politécnica de Calaluña, Barcelona, Spain, 2007.

52. Núñez, A.; Rueda, J.; Mezcua, J. A site amplification factor map of the Iberian Peninsula and the Balearic Islands. *Nat. Hazards* **2013**, *65*, 461–476. [CrossRef]

53. Álvarez-Cabal, R.; Díaz-Pavón, E.; Rodríguez-Escribano, R. *El terremoto de Lorca: Efectos en los edificios*, 1st ed.; Consorcio de Compensación de Seguros. Madrid, Spain, 2014.

54. López, M.F.P. *Riesgos catastróficos y ordenación del territorio en Andalucía, Consejería de Obras Públicas y Transportes*; Dirección General de Ordenación del Territorio y Urbanismo: Sevilla, Spain, 1999.

55. Goded, T.; Irizarry, J.; Buforn, E. Vulnerability and risk analysis of monuments in Málaga city's historical centre (Southern Spain). *Bull. Earthq. Eng.* **2012**, *10*, 839–861. [CrossRef]

56. Jaramillo-Morilla, A.; Mascort-Albea, E.J.; Ruiz-Jaramillo, J. Seismic microzoning maps for restoration of monuments in Seville: Nuestra Señora de los Reyes Convent. In Proceedings of the International Conference on Preservation, Maintenance and Rehabilitation of Historical Buildings Structures, Tomar, Portugal, 19–21 June 2014; pp. 937–943.

57. Ruiz-Jaramillo, J.; Mascort-Albea, E.J.; Jaramillo-Morilla, A. Proposed methodology for measurement, survey and assessment of vertical deformation of structures. *Struct. Surv.* **2016**, *34*. [CrossRef]

58. Instituto Andaluz de Geofísica. *Sismos que han afectado al territorio Andaluz con intensidad máxima igual o superior a VII*; Instituto Universitario de Investigación Andaluz de Geofísica y Prevención de Desastres Sísmicos: Granada, Spain, 2003; Available online: http://iagpds.ugr.es/pages/informacion_divulgacion/sismos_superior_vii (accessed on 10 April 2019).

59. Benito, M.B.; Navarro, M.; Vidal, F.; Gaspar-Escribano, J.; García-Rodríguez, M.J.; Martínez-Solares, J.M. A new seismic hazard assessment in the region of Andalusia (Southern Spain). *Bull. Earthq. Eng.* **2010**, *8*, 739–766. [CrossRef]

buildings

MDPI

Article

Potential Seismic Damage Assessment of Residential Buildings in Imzouren City (Northern Morocco)

Seif-eddine Cherif [1,*], Mimoun Chourak [2], Mohamed Abed [3] and Abdelhalim Douiri [4]

[1] Département de Génie Civil, Faculté des Sciences et Techniques, Université Cadi Ayyad, 112 Boulevard Abdelkrim Al Khattabi, Marrakesh 40000, Morocco

[2] Ecole Nationale des Sciences Appliquées Oujda, Université Mohammed Premier, BP 669, Oujda 60000, Morocco; mchourak00@gmail.com

[3] Civil Engineering Department, Université de Blida, Route de Soumàa, BP 270, Blida 09000, Algeria; abedmed@yahoo.fr

[4] Urban Agency of Al Hoceima 80, Avenue Mohamed V, El Hoceima 32000, Morocco; douiri.abdelhalim@gmail.com

* Correspondence: seif.cherif.00@gmail.com; Tel.: +212-699929208

Received: 9 November 2018; Accepted: 6 December 2018; Published: 11 December 2018

Abstract: The main purpose of this study is to assess seismic risk and present earthquake loss scenarios for the city of Imzouren, in northern Morocco. An empirical approach was chosen to assess the seismic vulnerability of the existing buildings, using the Vulnerability Index Method (RISK-UE), and considering two earthquake scenarios (deterministic and probabilistic). Special concern was given to the seismic vulnerability in Imzouren since the 2004 earthquake (24 February, mw = 6.4) that struck the region and caused substantial damage. A site investigation was conducted in the city targeting more than 3000 residential buildings, which had been closely examined and catalogued to assess their seismic vulnerability. The results of the seismic risk assessment in the city are represented through damage to the buildings, harm to the population and economic loss. Generally, the results obtained from the deterministic approach are in agreement with the damage caused by the 2004 earthquake.

Keywords: earthquakes scenarios; vulnerability index; seismic risk; Imzouren

---

1. Introduction

There are numerous uncertainties in the process of obtaining earthquake loss scenarios which are mainly related to either the seismic source or the building inventory [1]. Several methods have provided credible results while taking into account these uncertainties, such as performance-based methods and reliability-based design approaches [2–4]. Concerning the source location, earthquake scenarios can rarely be associated with certainty to the rupture of well-identified faults, since in the past, numerous destructive earthquakes occurred on blind faults or on offshore faults. In case the source location is uncertain, a probabilistic scenario may arguably be preferable to a deterministic one [5]. Still, if the site location is near a seismic source, first order rupture effects on ground motion are expected, which are difficult to assess in probabilistic hazards analysis. The second obstacle is the difficulty of categorizing the built environment into well-defined typologies and construction periods. Sometimes, renovations and post-earthquake reinforcements get in the way of properly characterizing the structural nature of the buildings, and thus properly estimating their seismic vulnerability on a large scale [1].

In case of this study, numerous challenges including the aforementioned problems are present in Imzouren. The city is located in the Al Hoceima region, which is the most seismically active zone in Morocco [6–8]. In ten years period, the region was struck by two destructive earthquakes; the first one in 26 May 1994 of magnitude 5.9 mw and the second and strongest one in 24 February 2004 of magnitude 6.4 mw [9]. Both were shallow earthquakes (depth < 15 km), but they didn't cause a surface rupture of a clear tectonic origin even though the observed cracks in the region back in 2004 were interpreted as such [10]. The identification of the responsible faults for these earthquakes was very problematic; especially since the assumed location of the epicenter had evolved over time [7,11–18]. Imzouren suffered the greatest damage in the aftermath of the 2004 earthquake that struck the region, while Al Hoceima experienced significantly less damage, even though the two cities are equally distant from the seismic source. Many believe that the damage difference is due to the bad design of the structures and the position of Imzouren in a soil field [9,18,19], which amplified the ground shaking.

Regarding the buildings, they are mostly reinforced concrete moment frame structures with masonry infill walls, mainly because the city was founded in the second half of the 20th century. One of the main problems that the built environment suffers from is a lack of supervision during the construction period of the buildings. In fact, many owners live abroad (Europe) and have their dwellings constructed without inspection and without any respect to the seismic standard in the region. Additionally, since many earthquakes have struck the city in the recent past (1994, 2004), there have been reconstructions and reinforcements of the damaged and affected buildings, which makes the estimation of the seismic vulnerability difficult.

Imzouren (35°09' N, 3°52' W) is located in the province of Al Hoceima on the northern coast of Morocco. Almost as important as Al Hoceima, Imzouren has different structural characteristics. It stretches along 3 km of the left margin of the Oued Nekkor and occupies part of the recent alluvial plain and oldest terraces formed by conglomerates and sandstones of the Pliocene age [19]. The 2004 population census for the province of Al Hoceima reported that Imzouren counts 26,474 inhabitants, 5147 residential buildings and an average of 5 inhabitants per dwelling [20]. As recently as the 1990s, constructive measures (BAEL 91 and PS92) [21,22] were introduced to protect buildings in the Rif region; the national seismic standard R.P.S. 2000 [23] came afterwards to set the essential seismic codes in the country [24]. According to the current Moroccan seismic code, the acceleration has a value of 0.18 g for an exceedance probability of 10% in a return period of 50 years [25].

The flow chart in Figure 1 shows an overview of the methodology applied in this paper. Seismic hazard assessment was evaluated in terms of macroseismic intensity, where probabilistic and deterministic scenarios were carried out. Site effects have also been considered in this study since Imzouren is formed on soft sediment, and were given incremental values added to the macroseismic intensity for both scenarios. The seismic vulnerability of the built environment was evaluated in terms of a Vulnerability Index Method (VIM) adapted and applied to the regional building characteristics. The seismic risk is represented by direct damage to the buildings, damage to population and economic loss. Ideally, earthquake loss models should include all possible induced phenomena from earthquakes: landslides, liquefaction, surface fault rupture, and tsunamis. However, strong ground motion is often the only hazard considered in loss assessment methods. It is commonly an acceptable approach because as the size of the loss model increases, the relative influence of the secondary hazards such as liquefaction and landslides decreases [26]. Harm to the population is defined in terms of casualties and people needing to be relocated (homeless), while economic loss is calculated based on reconstruction costs. All of these aspects are directly related to the direct damage on buildings.

The vast amount of building inventory data made the use of a Geographical Information System necessary for this study. Furthermore, the GIS-generated damage distribution maps also have the advantage of being easily understood and used by city planners and risk managers.

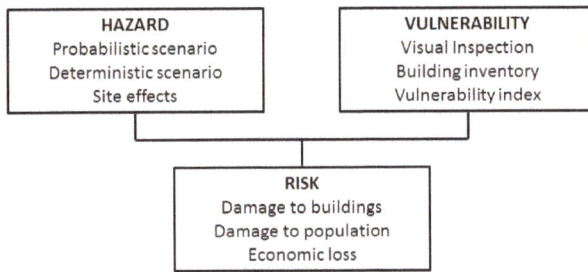

Figure 1. Flow chart of the adopted methodology for the seismic risk assessment in Imzouren.

2. Seismic Intensity and Earthquake Scenarios

2.1. Earthquake Scenarios

Two main scenarios were considered for this study. A deterministic scenario may be the better approach for a city in a seismic zone; however, we have also considered a probabilistic scenario to compare the differences between the two damage loss scenarios. Given the small size of Imzouren, the macroseismic intensity is considered to be constant all over the city.

The deterministic scenario is based on a reference event, which represents the closest earthquake to the target site that caused the greatest damage. In this case, we assume this event to be similar to the earthquake that struck the region of Al Hoceima in 24 February 2004 [9]. The earthquake took place at an epicentral distance of 12 km from the city and was estimated at a depth between 6 and 10 km. The damage caused was catastrophic; 629 dead, 966 injured and 15,600 homeless people [17]. The estimated intensity in the city of Imzouren was in the range of IX-X degree of MSK scale [27]. The same reference event was considered beforehand in the study of seismic risk in Al Hoceima [24] since the two cities are relatively close to each other.

The seismic hazard for the probabilistic scenario was addressed by the Risk Management Solutions Inc. in 2012 [28]. According to the results, the city of Imzouren was assigned an intensity of VIII in MSK scale [29] and a peak ground acceleration equivalent to 0.303 g for a return period of 475 years. According to the national seismic standard R.P.S. 2000, Version 2011 [25], the seismic acceleration for a return period of 475 years is equal to 0.18 g, which is significantly lower than the one estimated by the RMSI report [28].

2.2. Site Effects

The surface ground motion may be strongly amplified if the geological conditions are unfavorable; whether it's topography or surface failure or sedimentary basins. These geological specificities can strongly influence the nature and severity of shaking at a given site. For this study, site effects were estimated based on a seismic microzonation conducted in the city [30]. An iso-frequency map was elaborated using the Nakamura method H/V (Figure 2), which consists in estimating the ratio between the Fourier amplitude spectra of the horizontal (H) to vertical (V) components of ambient noise vibrations. The obtained spatial distribution map of iso-frequency values shows the existence of homogeneous zones that correlate well with the lithology of this area [30]. Three zones were considered: zone 1 with frequencies ranging from 0.93 Hz to 1.57 Hz (deposits of conglomerates, sandstone, coastal glaze loam and silt), zone 2 with frequencies between 1.57 Hz and 1.88 Hz (conglomerates, sandstones and silt trays and gray silt of the plain of Oued Nekor) and zone 3 with frequencies between 1.88 Hz and 4.90 Hz (schist bedrock, shale formations and rock).

Geotechnical characterization of the area around the town was carried out to define site effects in terms of intensity values, based on the spatial distribution map of soil frequencies (Figure 2). The intensity has been incremented by frequency margins and the used increments were decided based on expert opinion. The introduction of soil effects in terms of intensity was also recommended in the

RISK-UE project [31]. In zones 1 and 2, increments of 1 and 0.5 were applied respectively, while no intensity increments were considered for zone 3. The intensity map for the city of Imzouren, including the soil effects, is shown in Figure 3a,b according to both scenarios. As can be seen, the seismic intensity according to the deterministic scenario is significantly higher than the one estimated by the probabilistic scenario.

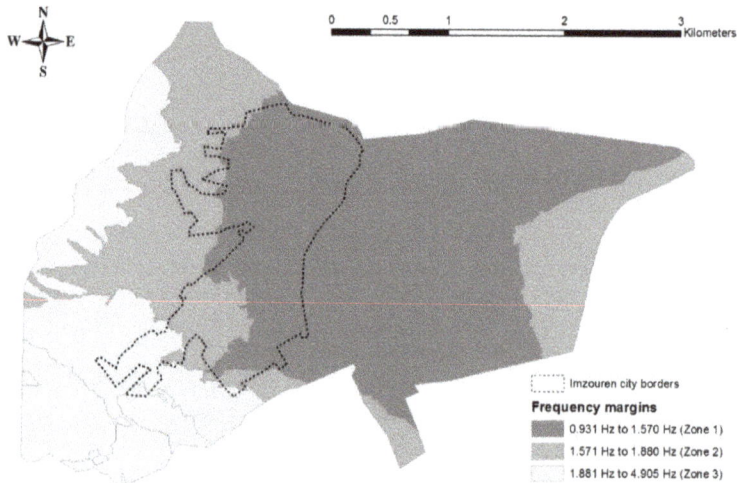

Figure 2. Spatial distribution of frequencies and identification of homogeneous areas in the town of Imzouren [30].

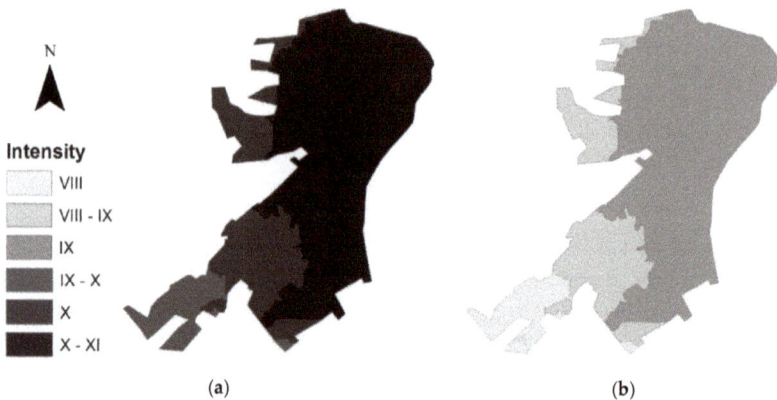

Figure 3. Intensity map for the city of Imzouren including soil effects according to (**a**) the deterministic scenario and (**b**) the probabilistic scenario.

3. Description and Classification of Imzouren Building Inventory

Imzouren is a new city where residential buildings are mostly low-rise reinforced concrete moment frame structures. According to the 2004 census [20] the modern Moroccan house is the predominant type, constituting 94% of total dwellings (Table 1). It is a city where the money from emigration to Europe was recently invested in multiple buildings of 3, 4 or 5 floors [19], and this is indicated in the results of the same census, where more than 30% of the dwellings are either vacant or seasonal

(Table 2). Most of the buildings have simple geometrical shapes with small construction areas ranging from 100 m^2 to 150 m^2 (Figure 4).

Table 1. Types of housing in the city of Imzouren [20].

Housing Types	Number	%
Villa	9	0.2
Apartment	4	0.1
Traditional Moroccan house	74	1.4
Modern Moroccan house	4834	93.9
Slum	20	0.4
Rural dwelling	23	0.4
Others	183	3.6

Table 2. Housing occupation in the city of Imzouren [20].

Housing Occupation	Number	%
Total dwellings	7471	100
Occupied dwellings	5122	68.6
Vacant dwellings	1492	20
secondary or seasonal dwellings	857	11.5

Figure 4. Modern Moroccan houses in the city of Imzouren.

For the purposes of this study, the city was subdivided into multiple sections representing 11 districts (Table 3). Each section is represented by a number of studied buildings that characterize the structural nature and the geometry of the residential buildings in the area. The investigation targeted buildings in the whole city, where geometrical features were inspected; from number of floors to irregularities and maintenance. A total of 3077 residential buildings spread throughout the city were the object of this study (Figure 5), which represents approximately 60% of the total number of residential buildings. It is also of interest to comment on the number of reinforcements and reconstructions that has been seen throughout the investigation, especially in the southern part of the city, where the damage caused by the 2004 earthquake was more important (Figure 5). However, the reinforcements aren't applied in the most efficient way; it mostly consists of additional steel bars to the first floor columns.

Table 3. Adopted code level for the existing buildings in Imzouren.

Construction Period	Before 1960	1960–1994	After 1994
Code level	Low code	Medium code	High code

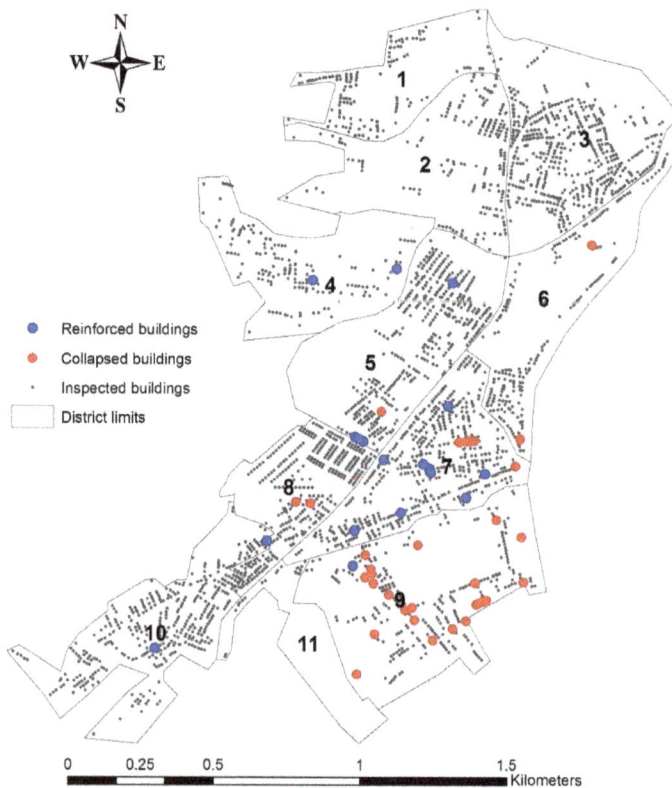

Figure 5. District limits and studied buildings in the city of Imzouren.

During the investigation, several obstacles were encountered, which will inevitably induce uncertainties in the results. Among the problems encountered, there were poor construction procedures, usually due to the lack of supervision and control on site, especially the linking of structural elements and the quality of construction materials, which can be very difficult to trace during the visual inspection. Also, the buildings share structural and architectural similarities, which is why it has proved difficult to characterize them individually. The code level was introduced in order to assess the seismic vulnerability of buildings more efficiently.

Three code levels were defined for residential buildings in Imzouren (Table 3), depending on the construction period; before 1960, between 1960 and 1994 and after 1994. The 2 events (1960, 1994) are very important in the seismic history of Morocco, given the fact that they represent a substantial change in construction habits, especially in the region of Al Hoceima [24]:

- The 1960 event: Agadir was struck by one of the most destructive earthquakes in the 20th century [32] on 29 February 1960, causing more than 12,000 fatalities. A first seismic standard resulted from the studies and investigations in site, named "Agadir Standard". Without a proper seismic code, the Moroccan construction has been greatly affected by this standard.
- The 1994 event: A violent earthquake ($M_w = 6.0$) struck the region of Al Hoceima, causing casualties and extensive damage [28,33]. Constructive measures have been taken post-earthquake to protect and reinforce buildings in the Rif region, based on the PS92 [22] and BAEL 91 [21]. These decisions were applied after 1994 in Al Hoceima, Imzouren and surrounding towns and, subsequently, contributed to developing the first national seismic code known as R.P.S. 2000 [23].

The distribution of the existing buildings according to number of stories and code level is shown in Figure 6, where most low-code constructions are only one or two-story buildings.

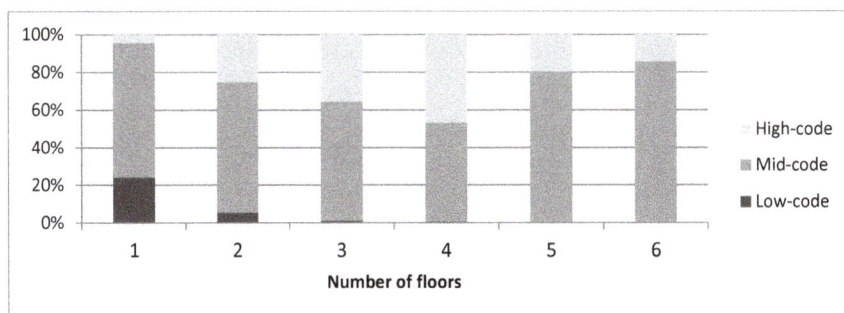

Figure 6. Distribution of the existing buildings according to number of stories and code level.

4. Seismic Vulnerability using the Vulnerability Index Method

The "Vulnerability Index Method" [34,35] has been widely used in Italy during the last decades. This method is considered "indirect" because the relationship between the seismic action and the seismic response is established by a "vulnerability index". The method uses a field survey form to gather information about the parameters of the buildings that might influence their vulnerability. The data on registered earthquakes is used to calibrate and adjust the vulnerability functions to link the vulnerability index (VI) to the main damage factor (d) for a specific seismic intensity or PGA.

RISK-UE (An Advanced Approach to Earthquake Risk Scenarios with Application to Different European Towns) was an important research project funded by the European Commission. Seven cities (Barcelona, Bitola, Bucharest, Catania, Nice, Sofia and Thessaloniki) were involved in this project, whose main objective was to develop a general methodology for assessing seismic risk in European cities [36–39]. The Vulnerability Index Method was chosen as one of the vulnerability assessment procedures that have been successfully developed and applied to all the cities mentioned above.

The main advantage of the "Vulnerability Index" methods is that they can determine the vulnerability characteristics of each building, rather than defining the vulnerability based only on the typology. However, the method requires expert judgment since the coefficients and weights applied in calculating the vulnerability index have a degree of uncertainty that is not taken into account. In addition, the calculation of the vulnerability index for a large building stock would be very time consuming, if such data is not already available [40].

The Vulnerability Index Method applied in the RISK-UE project was used for the purposes of this study. This method provides a typological classification system [41], to group structures with the same seismic performance V_I^{class} and then adds the Vm_j behavior modifiers specific to each building, to calculate a total vulnerability index $V_I^{building}$ for each building, using the following Equation [37]:

$$V_I^{building} = V_I^{class} + \Delta M_R + \sum_{j=1}^{n} Vm_j \tag{1}$$

where ΔM_R is a regional modifier which takes into account the characteristics of the region or the building period. The total vulnerability index $V_I^{building}$ takes values ranging from 0 (least vulnerable building) to 1 (most vulnerable building).

The method was adapted to the Moroccan features of the buildings [24] and applied to the studied buildings in Imzouren. The results show that the vulnerability index takes values ranging from 0.2 to 0.86, with an average value of 0.38 (Figure 7). The city has a low vulnerability, as can be seen from Table 4. The mean vulnerability indices for the different districts have values ranging between

0.31 and 0.44. This could be related to the reconstructions and reinforcements of the buildings that are accounted for since the 2004 earthquake. However, the existence of a minority of buildings can't be overlooked (7.5% of the total number of the studied structures) having a vulnerability index greater than 0.5 (Figure 7) and can be exposed to the collapse in case of an earthquake. The results correlate well with the 2004 census, where 4.7% of the households are below the relative poverty line and 7.6% are below the vulnerability threshold [20].

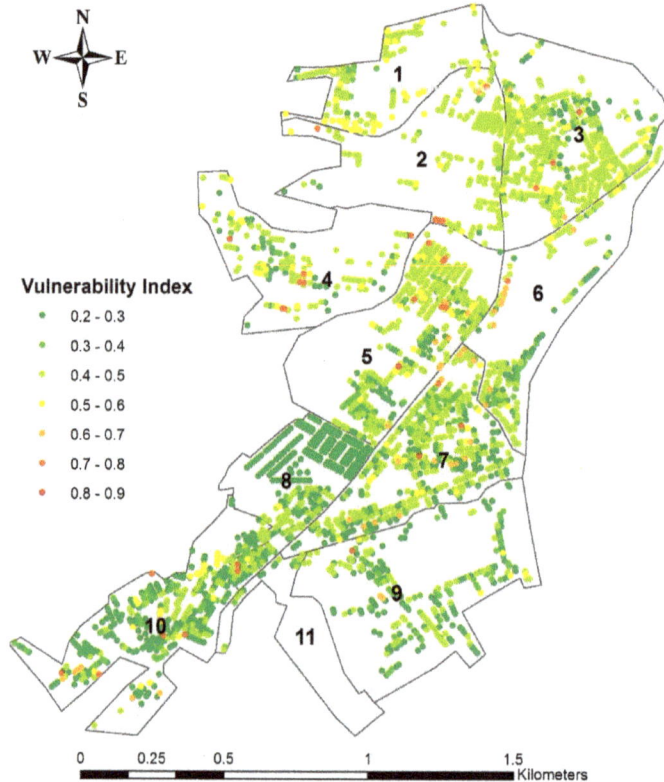

Figure 7. Vulnerability index values for the existing buildings in Imzouren.

Table 4. Mean Vulnerability Indices (MVI) for the Districts of Imzouren.

Label	District Name	Number of Inspected Buildings	MVI
1	Laazib	131	0.44
2	Iboujiren	185	0.44
3	Zaouia	530	0.42
4	Ait Moussa et Amar	155	0.41
5	Quartier Commercial	396	0.41
6	Ait M'hand Ou Yahya	182	0.40
7	Quartier Masjid	478	0.38
8	Souk	339	0.31
9	Tanaouia 1	267	0.33
10	Hay Rabia	392	0.32
11	Tanaouia 2	22	0.32

The VIM introduces five non-null damage states; Slight, Moderate, Substantial to Heavy, Very Heavy and Destruction [42]. The mean damage grade μ_D is introduced to characterize the likely damage to the building, for a given vulnerability (V_I) and a macroseismic intensity (I) according to the following Equation:

$$\mu_D = 2.5\left[1 + \tanh\left(\frac{I + 6.25V_I - 13.1}{\phi}\right)\right] \tag{2}$$

where ϕ is the ductility index, which is assessed taking into account the typology of the building and its geometrical and material characteristics [43]. For residential buildings, it has a value of 2.3 [36]. The distribution of the mean damage grade for the different building codes is shown in Figure 8. It is in fact the most important factor in the seismic vulnerability assessment in the city, since all existing residential buildings have the same structural build (reinforced concrete moment frame structures) and the same height (low to mid rise).

A weighted average index of damage DS_m can be calculated using the following Equation [37]:

$$DS_m = \sum_{k=0}^{5} kP[DS_k] \tag{3}$$

where k represents the state of damage taking values from 0 to 5 and $P[DS_k]$ represents the corresponding probabilities of occurrence of the damage state k. The damage distribution is calculated using the beta distribution [37].

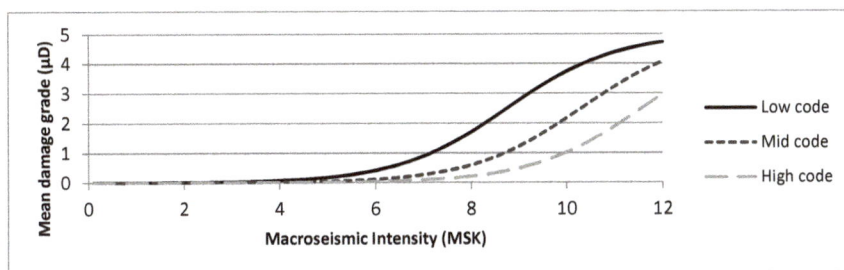

Figure 8. Mean damage grade for the considered building codes.

5. Seismic Risk and Loss Scenarios

The data gathered on the seismic vulnerability of the buildings was combined with the macroseismic intensity of the region, including site effects. The results are shown below according to a deterministic and a probabilistic scenario.

5.1. Direct Damage

The mean damage grade for each district can be seen in Figure 9 for both the deterministic and probabilistic hazard scenarios. The distribution of damage follows the same pattern as the site effects (Figure 3); the damage is more significant in the eastern part of the city where site effects are more present. The damage grade values range between 0.58 to 4.59 for the deterministic scenario and 0.17 to 3.77 for the probabilistic scenario, with mean values of 2.08 and 0.89 respectively. This corresponds to a moderate damage for the deterministic scenario and a slight damage for the probabilistic scenario. It is also important to note that the number of structures with a Very Heavy mean damage grade (partial collapse) is considerable for the deterministic scenario (73 structures).

Figures 10 and 11 display the distribution of damage in the different districts in the city according to the deterministic and probabilistic hazard scenarios. Similar to the mean damage factor (Figure 9), the damage distribution difference between the considered scenarios is significant, since the considered intensities affecting both scenarios are quite different (IX-X for the deterministic scenario and VIII for the probabilistic scenario).

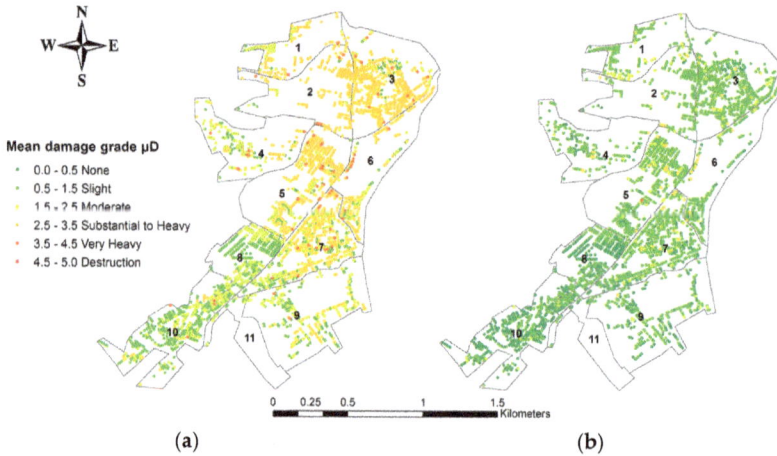

(a)   (b)

Figure 9. Mean damage grade for the residential buildings of Imzouren according to (a) the deterministic scenario and (b) the probabilistic scenario.

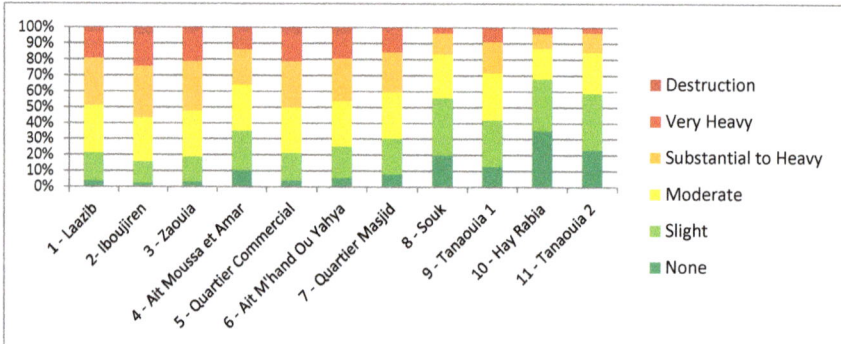

Figure 10. Damage distribution by districts according to the deterministic hazard scenario.

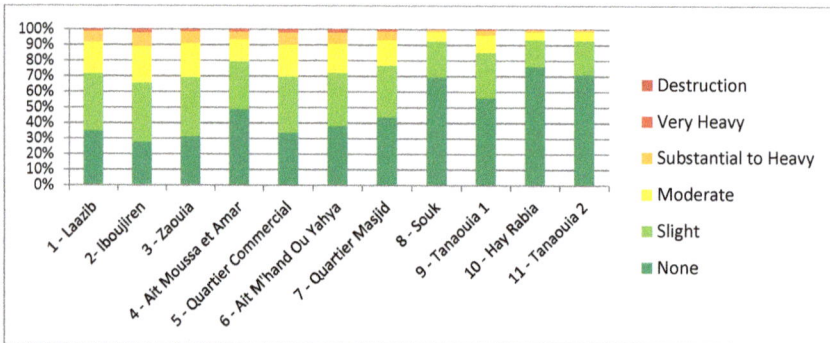

Figure 11. Damage distribution by districts according to the probabilistic hazard scenario.

5.2. Harm to Population

Harm to the population is estimated considering the number of casualties and homeless. First, to evaluate the number of fatalities and injured, the casualty model given by Coburn and Spence (2002) is used [32]:

$$K_S = C \cdot [M1 \cdot M2 \cdot M3 \cdot (M4 + M5 \cdot (1 - M4))] \tag{4}$$

The parameter K_S represents the number of casualties, C is the number of collapsed buildings, obtained by the number of buildings multiplied by the collapse probability. $M1$ is the number of inhabitants per building, according to the [20], an average of five persons per household are estimated for the city. $M2$ is the occupancy at time of the earthquake and a value of 75% for residential buildings is assumed for this case. $M3$ is the percentage of occupants trapped by collapse. Since Imzouren is a case of residential low-rise buildings, a percentage of 50% is assumed. $M4$ is the estimated injury distribution at collapse and the parameter depends on the building typology. Multiple cases are considered for the $M4$ parameter (Table 5): light injuries not necessitating hospitalization, injuries requiring hospital treatment, life-threatening cases needing immediate medical attention and dead or unsaveable cases. $M5$ is the percentage of trapped survivors in collapsed buildings that subsequently die, assumed to be equal to 90%. Table 5 shows the adopted values of $M2$–$M5$ parameters.

Table 5. The assumed values for the M2–M5 parameters [32].

Typology	M2 (%)	M3 (%)	M4 (%)				M5 (%)
			Light Injuries	Injuries Requiring Hospitalization	Life-Threatening Cases	Fatalities	
Values for reinforced concrete buildings	75	50	10	40	10	40	90

The results of casualties are carried out only for the deterministic scenario, where the case of collapsed buildings is possible (Figures 10 and 11). Table 6 shows the total injuries and fatalities for each district in the city of Imzouren. The estimated number of casualties is 580 and the number of fatalities is estimated at 147, which is strongly correlated with the estimated number of collapses that is equal to 84 for the deterministic scenario.

Table 6. Number of casualties for the deterministic scenario for each district in Imzouren.

Label	District Name	Light Injuries	Injuries Requiring Hospitalization	Life-Threatening Cases	Fatalities
1	Laazib	12	13	12	13
2	Iboujiren	36	37	36	37
3	Zaouia	31	32	31	32
4	Ait Moussa et Amar	7	7	7	7
5	Quartier Commercial	26	27	26	27
6	Ait M'hand Ou Yahya	15	16	15	16
7	Quartier Masjid	10	10	10	10
8	Souk	1	1	1	1
9	Tanaouia 1	5	5	5	5
10	Hay Rabia	0	0	0	0
11	Tanaouia 2	0	0	0	0
	Total	143	147	143	147

Harm to population was afterwards estimated in terms of the number of homeless people. The number of persons to be relocated because of uninhabitable buildings is also an important parameter in disaster management. The methodology that was applied is based on HAZUS 1999 [44] and considers that 100% of partially or completely destroyed buildings and 90% of the buildings with Heavy damage are considered uninhabitable.

The total number of residential units uninhabitable because of structural damage is calculated by the following Equation:

$$\%MF = 0.9 \times \%H_{MF} + 1.0 \times \%VH_{MF} + 1.0 \times \%D_{MF}$$
$$UNU_{SD} = U_{MF} \times \%MF \tag{5}$$

where U_{MF} is the total number of multi-family residential units, $\%H_{MF}$, $\%VH_{MF}$ and $\%D_{MF}$ are the probabilities corresponding to Substantial to Heavy damage, Very Heavy damage and collapse states respectively.

The estimated number of homeless people according to both deterministic and probabilistic scenarios is shown in Figure 12. The total number of homeless people estimated in both scenarios is 6404 and 221 respectively. The big difference between the two results is due to the fact that the damage exceeded the "Substantial to Heavy" limit in the deterministic scenario.

(a)   (b)

Figure 12. Number of homeless people for each district according to (a) the deterministic scenario and (b) the probabilistic scenario.

5.3. Economic Cost

Economic losses are considered as the present costs of reconstructing the damaged buildings. This value is estimated by the cost of reconstruction of reinforced concrete buildings without including the cost of land. Absolute economic cost S_{Cost} in millions of euros is given by the following equation [44]:

$$S_{Cost} = \sum_{k=2}^{5} CS(k) = V_C \sum_{K=2}^{5} \sum_{J=1}^{Ne} [Area(j) \cdot P_s(k,j) \cdot RC(k,j)] \tag{6}$$

where S_{Cost} is the sum of repair costs $CS(k)$ due to the damage state k; V_C is the cost per unit area. A constant value of V_C is assumed for all types of buildings [24]; Area is the building area; $P_s(k,j)$ is the probability for the construction j to be in the state of damage k and $RC(k,j)$ is the value of compensation due to the degree of damage k to the building j and is given as a percentage of the reconstruction cost per square meter. The used cost estimation doesn't take into account other factors such as finance charges, resale and residual values, or repair costs, all considered in the LCCA method for a single building or a building system [45]. However, given the objectives of the study and the scale of work, the economic estimate is considered satisfactory.

Figure 13 shows the economic losses in millions of euros for each section of the city, caused by the deterministic and probabilistic hazard scenarios. The total economic cost of the city is 197 Million

Euros in the case of the deterministic scenario, and 48.4 million euros in the case of the probabilistic scenario. The losses according to the deterministic scenario represent 43% of the estimated losses of the 2004 earthquake, taking into account the overtime dollar value and the exchange rates [46], which is fairly accurate given the statements in the technical reports of the overall damage distribution in the region of Al Hoceima [9,20,47].

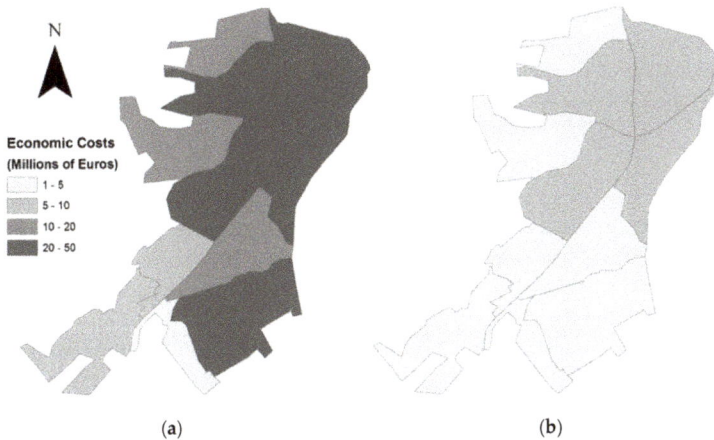

(a)   (b)

Figure 13. Distribution of economic cost according to (**a**) the deterministic scenario and (**b**) the probabilistic scenario.

6. Discussion

The results show that most buildings have a low vulnerability index, which can be linked to the improved quality of construction materials and construction practices since the 2004 earthquake. In addition, an increasing number of reinforcements and reconstructions have been identified throughout the city since the 2004 event. However, this fact doesn't exclude 7.5% of the buildings having important vulnerability indices, which is mainly related to the failure to comply to seismic construction regulations.

The results also show that direct damage in the city is moderate for the deterministic scenario, and light for the probabilistic scenario. The damage difference between the two considered scenarios is rather important, since the considered intensities affected to both scenarios are quite different. The results show similarities with the conducted surveys of the reference earthquake. According to damage surveys of the 2004 earthquake, buildings belonging to the same vulnerability class (B and C from EMS 98) show different results, some stood in good condition while others (around 30) completely collapsed [19,47]. All the reports believe the damage is related to the site effects and construction defects.

The estimated number of casualties is significant in case of the deterministic scenario, while inexistent for the probabilistic scenario. A total number of 84 Collapses, 580 casualties and 147 fatalities are estimated in the worst-case scenario. According to NatCatSERVICE, the overall losses due to the 2004 earthquake in the region of Al Hoceima is estimated around 400 million dollars [46]. The economic losses in the city of Imzouren according to the deterministic scenario are about 197 million euros and represent 43% of the estimated overall losses of the 2004 earthquake, taking into account the overtime dollar value and the exchange rates, which is a good estimate.

During the investigation, it was rather difficult to distinguish buildings by their characteristics since they have the same structural build and the same appearance. However, the existing dwellings were mainly categorized by their code level, in other words, whether the buildings respect new and old seismic construction standards or not. Additionally, this study doesn't take into account the poor construction practices that are not apparent on buildings (e.g., quality of construction materials, column-beam connections) which frequently occur when there is lack of supervision.

Based on the estimated states of damage according to both deterministic and probabilistic scenarios in the city of Imzouren, it is important to consider risk mitigation actions, which may include:

- Control of the mechanical characteristics of the existing buildings suspects of construction defects;
- Planning a seismic microzonation of the city using Vs_{30};
- Updating the seismic standard in the region of Al Hoceima, by adding response spectra for the cities and fragility curves for the typical structures;
- Increasing the level of preparedness of disaster and rising societal awareness of seismic risk.

7. Conclusions

Seismic risk in the city of Imzouren was assessed with an empirical approach based on site investigations and vulnerability index scoring. The evaluation of the seismic hazard and seismic vulnerability allowed a proper estimation of damage based on two seismic hazard scenarios; a deterministic scenario and a probabilistic one. The seismic hazard was defined in terms of macroseismic intensity including site effects.

A building inventory of 3077 residential buildings was collected for the purposes of this study. The inspection results indicate that all buildings are reinforced concrete moment frame constructions with masonry infill walls. The seismic vulnerability of the buildings was performed using the Vulnerability Index Method applied in the RISK-UE project. The results show that most buildings have a low vulnerability index. However, this fact doesn't exclude 7.5% of the buildings having important vulnerability indices. In terms of seismic risk results, direct damage in the city is considered to be moderate for the deterministic scenario, and light for the probabilistic scenario. The same difference in scenarios is observed for the harm to the population and economic costs, since these aspects are tightly linked to direct damage.

Author Contributions: Investigation, Conceptualization, Writing-Original Draft Preparation, S.e.-C.; Editing, Supervision, Project Administration, M.C.; Supervision, Data Curation, M.A.; Validation, Data Acquisition, A.D.

Funding: This research received no external funding.

Conflicts of Interest: The authors declare no conflict of interest.

References

1. Faccioli, E.; Pessina, V.; Calvi, G.M.; Borzi, B. A Study on Damage Scenarios for Residential Buildings in Catania City. *J. Seismol.* **1999**, *3*, 327–343. [CrossRef]
2. Bertero, R.D.; Bertero, V.V. Performance-based seismic engineering: The need for a reliable conceptual comprehensive approach. *Earthq. Eng. Struct. Dyn.* **2002**, *31*, 627–652. [CrossRef]
3. Zou, X.K.; Wang, Q.; Li, G.; Chan, C.M. Integrated reliability-based seismic drift design optimization of base-isolated concrete buildings. *J. Struct. Eng.* **2010**, *136*, 1282–1295. [CrossRef]
4. Castaldo, P.; Palazzo, B.; Ferrentino, T. Seismic reliability-based ductility demand evaluation for inelastic base-isolated structures with friction pendulum devices. *Earthq. Eng. Struct. Dyn.* **2017**, *46*, 1245–1266. [CrossRef]
5. McGuire, R. Scenario earthquakes for loss studies based on risk analysis. In Proceedings of the 5th International Conference on Seismic Zonation, Nice, France, 17–19 October 1995; pp. 1325–1333.
6. Cherkaoui, T.; Hatzfeld, D.; Jebli, H.; Medina, F.; Caillot, V. Etude microsismique de la région d'Al Hoceima. *Bull. Inst. Sci. Rabat* **1990**, *4*, 25–34.
7. El Alami, S.O.; Tadili, B.A.; Cherkaoui, T.E.; Medina, F.; Ramdani, M.; Ait Brahim, L.; Harnafi, M. The Al Hoceima earthquake of May 26, 1994 and its aftershocks: A seismotectonic study. *Ann. Geophys.* **1998**, *41*, 519–537.

8. Bezzeghoud, M.; Buforn, E. Source parameters of the 1992 Melilla (Spain, Mw = 4.8), 1994 Al Hoceima (Morocco Mw = 5.8) and Mascara (Algeria, Mw = 5.7) earthquakes and seismotectonic implications. *Bull. Seismol. Soc. Am.* **1999**, *89*, 359–372.

9. Jabour, N.; Kasmi, M.; Menzhi, M.; Birouk, A.; Hni, L.; Hahou, Y.; Timoulali, Y.; Bedrane, S. *The February 24th, 2004 Al Hoceima Earthquake*; Newsletter of European-Mediterranean Seismological Centre: Essonne, France, 2004.

10. Douiri, A. Thèse: Etude de L'aléa Sismique dans la Ville d'Imzouren: Effets de Site et la Vulnérabilité Sismique des Édifices. Ph.D. Thesis, FST Tanger, University Abdelmalek Essaâdi, Tétouan, Morocco, 2014.

11. Calvert, A.; Gomez, F.; Seber, D.; Barazangi, M.; Jabour, N.; Ait Brahim, L.; Demnati, A. An Integrate Geophysical Investigation of Recent Seismicity in the Al-Hoceima Region of North Morocco. *Bull. Seismol. Soc. Am.* **1997**, *87*, 637–651.

12. El Mouraouah, A.; Iben brahim, A.; Kasmi, M.; Birouk, A.; El M'rabet, E.M.; Zouine, R.; El Mouraouah, E.A.; Toto, M.; Benammi, M.; Hafid, S.; et al. *Le séisme d'Al Hoceima du 24 Février 2004. Surveillance Sismique et Analyse Préliminaire. Situation au 31 mai 2004*; CEPRIS-CNRST: Rabat, Morocco, 2004; 162p, ISBN 9981-41-007-01.

13. Hahou, Y.; Jabour, N.; Oukemeni, D.; El Wartiti, M.; Nakhcha, C. The Earthquake of 26 May 1994, Al Hoceima, Morocco; Intensity Distribution and Macroseismic Epicenter. *Seismol. Res. Lett.* **2004**, *75*, 51–58. [CrossRef]

14. Stich, D.; Mancilla, F.; Baumont, D.; Morales, J. Source analysis of the *Mw* 6.3 2004 Al Hoceima earthquake (Morocco) using regional apparent source time functions. *J. Geophys. Res.* **2005**, *110*. [CrossRef]

15. Biggs, J.; Bergman, E.; Emmerson, B.; Funning, G.; Jackson, J.; Parsons, B.; Wright, T. Fault Identification for Buried Strike-Slip Earthquakes using InSAR: The 1994 and 2004 Al Hoceima, Morocco Earthquakes. *Geophys. J. Int.* **2006**, *166*, 1347–1362. [CrossRef]

16. Cakir, Z.; Meghraoui, M.; Akoglu, A.M.; Jabour, N.; Belabbes, S.; Ait-Brahim, L. Surface Deformation Associated with the Mw 6.4, 24 February 2004 Al Hoceima, Morocco, Earthquake Deduced from InSAR: Implications for the Active Tectonics along North Africa. *Bull. Seismol. Soc. Am.* **2006**, *96*, 59–68. [CrossRef]

17. Tahayt, A.; Feigl, K.L.; Mourabit, T.L.; Rigo, A.; Reilinger, R.; McClusky, S.; Fadil, A.; Berthier, E.; Dorbath, L.; Serroukh, M.; et al. The Al Hoceima (Morocco) earthquake of 24 February 2004, analysis and interpretation of data from ENVISAT ASAR and SPOT5 validated by ground-based observations. *Remote Sens. Environ.* **2009**, *113*, 306–316. [CrossRef]

18. Van Der Woerd, J.; Dorbath, C.; Ousadou, F.; Dorbath, L.; Delouis, B.; Jacques, E.; Tapponnier, P.; Hahou, Y.; Menzhi, M.; Frogneux, M.; et al. The Al Hoceima Mw 6.4 earthquake of 24 February 2004 and its aftershocks sequence. *J. Geodyn.* **2014**, *77*, 89–109. [CrossRef]

19. Goula, X.; Gonzalez, M. *Visite Technique à la Zone Endommagé par le Séisme d'Al-Hoceima du 24 Février 2004*; ICC-GS-192/2004 (fr), Institut Cartogra'fic de Catalunya, Generalitat de Catalunya: Barcelona, Spain, 2004.

20. RGPH. Recensement Général de la Population et de L'habitat 2004. Available online: http://www.hcp.ma/recensement-general-de-la-population-et-de-l-habitat-2004_a633.html (accessed on 1 September 2018).

21. BAEL 91. *Règles Techniques de Conception et de Calcul des Ouvrages et Constructions en Béton Armé Suivant la Méthode des États-Limites*; Editions Eyrolles: Paris, France, 1992.

22. PS92. Règles PS Applicables Aux Bâtiments; NF P 06-013-DTU Règles PS 92. December 1995. Available online: https://www.icab.fr/guide/ps92/ (accessed on 10 December 2018).

23. R.P.S. 2000. *Règlement de Construction Parasismique. Ministère de l'Aménagement du Territoire, de l'Urbanisme, de l'Habitat et de l'Environnement*; Secrétariat d'Etat à l'Habitat: Rabat, Morocco, 2002.

24. Cherif, S.E.; Chourak, M.; Abed, M.; Pujades, L. Seismic risk in the city of Al Hoceima (north of Morocco) using the vulnerability index method, applied in Risk-UE project. *Nat. Hazards* **2017**, *85*, 329–347. [CrossRef]

25. R.P.S. 2000. *Version 2011 Règlement de Construction Parasismique*; Ministère de l'Habitat, de l'Urbanisme et de l'Aménagement de l'Espace: Rabat, Morocco, 2011.

26. Bird, J.F.; Bommer, J.J. Earthquake Losses due to Ground Failure. *Eng. Geol.* **2004**, *75*, 147–179. [CrossRef]

27. Talhaoui, A. Risques Géologiques et Activité Sismique Dans la Région d'Al Hoceima: Impact sur L'environnement et L'aménagement. Ph.D. Thesis, Faculty of Sciences University Rabat, Rabat, Morocco, 2005; 207p.

28. RMSI. *Morocco Natural Hazards Probabilistic Analysis and National Strategy Development*; Earthquake Hazard Report; Department of Economic and General Affairs: Rabat, Morocco, 2012.

29. Medvedev, S.V.; Sponheuer, W.; Karnik, V. *Seismic Intensity Scale Version MSK 1964*; United Nation Educational, Scientific and Cultural Organization: Paris, France, 1965; p. 7.
30. Douiri, A.; Mourabit, T.; Cheddadi, A.; Chourak, M. Pertinence de la méthode H/V «bruit de fond» pour l'évaluation de l'effet de site dans la ville d'Imzouren (Rif central, Maroc). *Notes Mém. Serv. Géol.* **2014**, *577*, 1–6.
31. Faccioli, E. Seismic hazard assessment for derivation of earthquake scenarios in Risk-UE. *Bull. Earthq. Eng.* **2006**, *4*, 341–364. [CrossRef]
32. Coburn, A.; Spence, R. *Earthquake Protection*, 2nd ed.; Wiley: Chichester, UK, 2002.
33. Akoglu, A.M.; Cakir, Z.; Meghraoui, M.; Belabbes, S.; El Alami, S.O.; Ergintav, S.; Akyuz, H.S. The 1994–2004 Al Hoceima (Morocco) earthquake sequence: Conjugate fault ruptures deduced from InSAR. *Earth Planet Sci. Lett.* **2006**, *252*, 467–480. [CrossRef]
34. Benedetti, D.; Petrini, V. Sulla Vulnerabilità Di Edifici in Muratura: Proposta Di Un Metodo Di Valutazione. *L'industria Delle Costr.* **1984**, *149*, 66–74.
35. GNDT. Seismic Risk of Public Buildings Part 1—Methodology Aspects. GNDT-CNR Report. 1993. Available online: https://emidius.mi.ingv.it/GNDT2/Pubblicazioni/Monografie_disponibili.htm (accessed on 11 December 2018). (In Italian)
36. Lantada, N.; Irizarry, J.; Barbat, A.H.; Goula, X.; Roca, A.; Susagna, T.; Pujades, L.G. Seismic hazard and risk scenarios for Barcelona, Spain, using the Risk-UE vulnerability index method. *Bull. Earthq. Eng.* **2010**, *8*, 201–229. [CrossRef]
37. Milutinović, Z.V.; Trendafiloski, G.S. *WP04: Vulnerability of Current Buildings Handbook. RISK-UE Project: An Advanced Approach to Earthquake Risk Scenarios with Applications to Different European Towns*; Contract No. EVK4-CT-2000-00014; Institute of Earthquake Engineering and Engineering Seismology (IZIIS): Skopje, Macedonia, 2003.
38. Mouroux, P.; Bertrand, M.; Bour, M.; Brun, B.L.; Depinois, S.; Masure, P.; Risk-UE Team. The European Risk-UE project: An advanced approach to earthquake risk scenarios. In Proceedings of the 13th World Conference Earthquake Engineering, Vancouver, BC, Canada, 1–6 August 2004.
39. Pitilakis, K.; Alexoudi, M.; Argyroudis, S.; Anastasiadis, A. Seismic risk scenarios for an efficient seismic risk management: The case of Thessaloniki (Greece). *Adv. Earthq. Eng. Urban Risk Reduct.* **2006**, *66*, 229–244. [CrossRef]
40. Calvi, G.M.; Pinho, R.; Magenes, G.; Bommer, J.J.; Restrepo-Vélez, L.F.; Crowley, H. Development of seismic vulnerability assessment methodologies over the past 30 years. *ISET J. Earthq. Technol.* **2006**, *43*, 75–104.
41. Giovinazzi, S.; Lagomarsino, S. *WP04: Guidelines for the Implementation of the I Level Methodology for the Vulnerability Assessment of Current Buildings*; Risk-UE Project: Genoa, Italy, 2002.
42. Grünthal, G. *European Macroseismic Scale 1998*; Centre Européen de Géodynamique et de Séismologie: Luxemburg, 1998.
43. Lagomarsino, S.; Giovinazzi, S. Macroseismic and mechanical models for the vulnerability and damage assessment of current buildings. *Bull. Earthq. Eng.* **2006**, *4*, 415–443. [CrossRef]
44. HAZUS. *Earthquake Loss Estimation Methodology Technical Manual*; National Institute of Building Sciences for Federal Emergency Management Agency (FEMA): Washington, DC, USA, 1999.
45. Fuller, S. *Life-Cycle Cost Analysis (LCCA). Whole Building Design Guide*; WBDG: Washington, DC, USA, 2010.
46. NatCatSERVICE. *Natural Disasters 2004–10 Deadliest Natural Disasters*; Munich Re, Geo Risks Research: Munich, Germany, 2008.
47. Cherkaoui, T.E.; Harnafi, M. *Le Séisme d'Al Hoceima du 24 Février 2004*; Travail Collectif National Sur: Région de Taza-Al Hoceima-Taounate, Ressources et Stratégies de Développement, Faculté Polydisciplinaire de Taza: Taza, Morocco, 2005; pp. 27–38.

buildings

MDPI

Article

The Influence of Geo-Hazard Effects on the Physical Vulnerability Assessment of the Built Heritage: An Application in a District of Naples

Nicola Chieffo [1,*] and Antonio Formisano [2]

[1] Department of Civil Engineering, Faculty of Engineering, Politehnica University of Timisoara, Traian Lalescu Street, 300223 Timisoara, Romania
[2] Department of Structures for Engineering and Architecture, School of Polytechnic and Basic Sciences, University of Naples "Federico II", P. le V. Tecchio 80, 80125 Napoli, Italy; antoform@unina.it
* Correspondence: nicola.chieffo@student.upt.ro

Received: 27 December 2018; Accepted: 16 January 2019; Published: 21 January 2019

Abstract: The proposed study aims at analysing a sub-urban sector in the historic centre of Qualiano, located in the province of Naples (Italy), in order to assess the seismic vulnerability of the main typology classes (masonry and reinforced concrete) in the study area and the consequent expected damage scenarios. The typological and structural characterisation of the investigated area is done through the CARTIS form developed by the PLINIVS research centre together with the Italian Civil Protection Department. Subsequently, the vulnerability simulation analysis is carried out by means of a quick methodology integrated into a GIS tool in order to identify the structural units (S.U.) most susceptible at damage under seismic events. Furthermore, in order to take into account the possible damage scenarios, a parametric analysis is performed using a seismic attenuation law in order to obtain the maximization of the expected urban losses. Finally, the site and topographical local conditions, which negatively influence the severity of the seismic damage on the structures, have been taken into account in order to more correctly foresee the expected damage of the inspected sub-urban sector to be used for appropriate seismic risk mitigation plans.

Keywords: CARTIS form; Vulnerability assessment; Seismic attenuation law; Expected damage scenario; GIS mapping; Geo-hazard site effects

1. Introduction

The seismic risk assessment of urbanized areas involves a multi-level analysis that is based on the combination of three main factors, such as, Hazard (H), Exposure (E), and Vulnerability (V) [1]. Within this approach devoted to the seismic risk assessment, the large-scale vulnerability of building samples assumes a noteworthy importance, as it is one of the main factors that is responsible for structural damages due to possible seismic events [2]. In fact, the rigorous assessment of the vulnerability of existing buildings and the implementation of appropriate solutions to mitigate the earthquake induced effects aims at reducing the levels of physical damage and socio-economic losses impact of the future seismic events. Risk management of urban areas is normally carried out without an adequate spatial planning tool. In fact, the high population density and the absence of adequate renovation interventions on buildings are factors that negatively influence the global vulnerability of entire urban centres with disastrous consequences in the case of seismic events [3]. As a result of this condition, the main cause of huge social losses, observed during the past earthquakes, is due to the collapse of buildings. However, the general concept of risk, independent from the considered natural phenomena, can be understood as the achievement of a level of "expected losses", social and economic, into a given area [4]. It is, therefore, important to know the urban heterogeneity and the global socio-economic level to guarantee the development of policies and programs that prevent and reduce urban vulnerability in relation to the number of inhabitants and their exposure to natural phenomena. In this context, it seems evident that the loss concept is commonly identified as costs that should be sustained in order to restore the system original configuration.

The assessment of large-scale seismic vulnerability, therefore, would allow for both estimating the potential future losses due to the occurrence of earthquakes that can affect a particular region and supporting the technicians to implement risk mitigation plans. This kind of approach was first proposed by the Italian National Group for the Defence against Earthquakes (GNDT), which took the profit of post-earthquake damage observations of masonry buildings in Italy [5]. This method considered buildings as isolated structures [6], thus neglecting the aggregated configuration, where the interactions and the connections among adjacent constructions modify their seismic behaviour. Therefore, this analysis approach has been subsequently modified in order to consider any mutual interactions of buildings placed in aggregate due to earthquakes [7]. This procedure was then developed for several case studies in Europe [8–10].

Observational models have the peculiarity to be easily implemented into the macroseismic method [11] for the evaluation of the expected damage in a generic urbanized area according to the European macroseismic scale EMS-98 [12].

In the large-scale risk assessment, it is very important to take into account the selection of seismogenic zones, which are commonly referred to a scenario analysis, in order to have a realistic distribution of site effects on the entire territory. The geo-morphological effects have an important role on the seismic vulnerability investigation, since they significantly influence the behaviour of buildings due to the local amplification effects, an intrinsic feature of the site of interest. Generally, the expected risk is represented in terms of macroseismic intensity and empirical correlations between the surface geology and the seismic intensity, based on the post-event observations, are proposed. In fact, simplified methodologies are implemented for the estimation of site effects [13,14] and their influence on the seismic vulnerability impact scenarios. These studies are mainly based on the macroseismic approach, since the macroseismic intensity is the main parameter linking the direct effects of the earthquake to the consequent damage. The methodology that was proposed by Giovinazzi and Lagomarsino [15] allows for estimating the macroseismic intensity increase as a function of amplification factors, defined for different class of buildings and soil conditions. Therefore, according to the macroseismic approach [12], the soil conditions are taken into consideration within the vulnerability index and the risk is then evaluated in terms of macroseismic intensity as compared to the rigid soil requirement.

Based on these premises, the present work proposes a large-scale vulnerability study of a sub-urban sector of the historical centre of Qualiano, in the district of Naples, focusing the attention on the geo-hazard effects that are induced by local site phenomena. This is due to the fact that Qualiano rises on deposits of *lapilli* characterising the ground, which is classified as a category C. More in detail, the study aims at estimating the amplification of macroseismic intensity and, subsequently, of the seismic vulnerability of the inspected area in order to correctly estimate the expected damage scenarios.

2. The Historical Centre of Qualiano

2.1. Historical News

Qualiano, Figure 1, is a municipality with 25.704 inhabitants in the metropolitan area of Naples in Campania. The name derives from Caloianum, as can be seen from the first complete list of all the houses in the kingdom of Naples. The historical information on the origin of Qualiano are corroborated by numerous archaeological discoveries that allowed to suppose, without doubts, the role and function of the Roman village of Collana, which was dominated first by Greek colonies and then by Roman people.

Recent archaeological finds also suggest that Qualiano was one of the most favourite centres of Roman patricians for the healthy climate and the flourishing vegetation. This was also narrated by the historian Titus Livio, who described the natural riches of the entire Agro-Giuglianese territory. In fact numerous remains of the Roman empire, such as marble statues, *opus reticulatum masonry*, consisting of diamond-shaped tuff bricks placed around a core of *opus caementicium* (the former concrete), *opus latericium* masonry, which is coarse-laid brickwork used to face a core of *opus caementicium*, coloured mosaic floors dated from the end of the 1st century BC to the beginning of the 1st century AD, huge clay vases for storing foodstuffs (wheat, barley, etc.), and a rectangular water-filled cistern with a vaulted roof [16], were found.

Already in the fourth century, Qualiano recorded the presence of the Samnite people, while the formation of an agricultural centre or village (fagus) dated back to the III century BC, with the presence of the Oscan-Samnite people. The highest importance and urban growth of Qualiano occurred in the fourth or fifth century AD during the period of Roman decadence. The historical periods from the Angevin-Aragonese kingdom to the Spanish one (1500 and 1600) affected the modest agglomeration of Qualiano, while from the urban point of view the second Bourbon period (1815–1860) had a considerable importance.

Currently, Qualiano belongs the giuglianese area with the municipalities of Villaricca, Calvizzano, Giugliano in Campania, Marano di Napoli, and Mugnano di Napoli.

| (a) | (b) |

Figure 1. (a) Geographical location of Qualiano in the Campania region of Italy; and, (b) street view of one of the main city square.

2.2. Typological and Structural Characterisation of Samples of Buildings: The CARTIS Form

The CARTIS form [17] was developed by the PLINIVS research centre of the University of Naples "Federico II" in collaboration with the Italian Civil Protection Department (DPC) during the ReLUIS 2014–2016 project—"*Development of a systematic methodology for the assessment of exposure on a territorial scale based on the typological characteristics/structural of buildings*".

The CARTIS form aimed at detecting the prevalent ordinary building typologies in municipal or sub-municipal areas, called urban sectors, characterised by typological and structural homogeneity. The form refers only to ordinary buildings, mainly houses (multi-storey buildings) made of masonry and reinforced concrete (RC). Therefore, from the typological characterisation, monumental buildings (religious buildings, historic buildings, etc.), special structures (industrial warehouses, shopping centres, etc.), and strategic ones (hospitals, schools, etc.), whose characteristics are not comparable to ordinary buildings, have been excluded. The preliminary phase requires the identification of the homogeneous urban sector, appropriately bounded on the map and progressively numbered with respect to other sectors [18].

The form is divided into four sections: section 0 for the identification of the municipality and the sectors identified therein; section 1 for the identification of each of the predominant typologies characterizing the generic sub-sector of the assigned municipality; section 2 for the identification of general characteristics of each typology of constructions; section 3 for the characterization of structural elements of all individuated construction typologies.

It is worth noting that the structural-typological characterisation of the built-up represents a useful tool for improving the inventory of building distributions on the national territory for the large scale seismic vulnerability assessment by means of any specific approach. The overlapping of the basic cartography with the elaborations related to the urban habitat chronological development or, in the absence of it, with the comparison between cadastral maps of different epochs, allow for knowing the phases of the city growth. From these elaborations, it is possible to identify the historical areas of Qualiano, which is those built before the seismic classification of the municipality (occurred after 1980) and the urbanised areas with recent buildings. Based on the aforementioned items, the municipality of Qualiano has been subdivided into four sectors. The first, C01, is the historical centre, the second C02, is the first urban expansion, the third, C03, is made of suburbs, and finally, the fourth, C04, is an agricultural/industrial area (Figure 2).

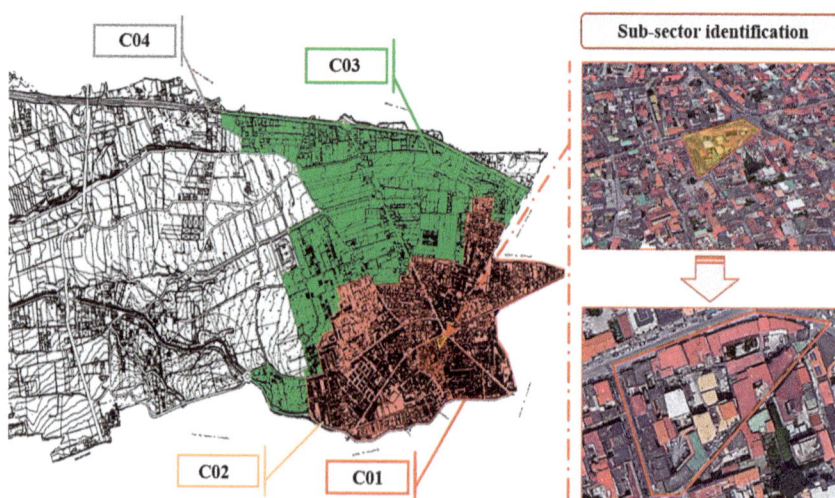

Figure 2. The main urban sectors of Qualiano.

The study is herein conducted on the C01 sector (Figure 3), where two types of masonry structures, URM1 and URM2, and one kind of reinforced concrete buildings, RC, are identified according to the acronyms of structural typologies given by the CARTIS form.

(a)   (b)

Figure 3. (a) Bird-eye view and (b) numbering of buildings in the selected sub-sector of the C01 sector.

The buildings of the URM1 typological class are placed in the heart of the historical centre of Qualiano, which collects all of the constructions built before 1860 and are generally composed of two floors.

From the technological point of view, these buildings have a tuff masonry structure with an average thickness of 0.80 m and direct masonry foundations. The lack of connections at corners among perimeter walls orthogonal to each other does not guarantee a global behaviour of the structure. The horizontal structures are characterised by timber floors and unusable roofs. Similarly, URM2 buildings have two floors above ground with inter-storey heights of 3.50 m. They have a tuff masonry structure with thicknesses of 0.80 m, reinforced concrete horizontal structures, and direct and continuous foundations. The arrangement of the openings on the façade is regular and buildings have a good conservation state.

Finally, the recently constructed reinforced concrete buildings consist of four floors with inverse beams foundations. Cladding walls are made of brick masonry with thickness of 0.30 m.

Subsequently, the acronyms proposed by the BTM (*Building Typology Matrix*) [19] have been used to indicate the typological classes of the CARTIS form, see Figure 4. In particular, the classes URM1 and URM2 correspond to M1.2 (tuff masonry stones with timber floors, 28%) and M3.4 (tuff masonry stones with reinforced concrete floors, 60%), respectively. Furthermore, RC1 is associated to the RC class (reinforced concrete frames, 12%). The representation on the Qualiano's map of the inspected building typological classes is done using the QGIS tool [20].

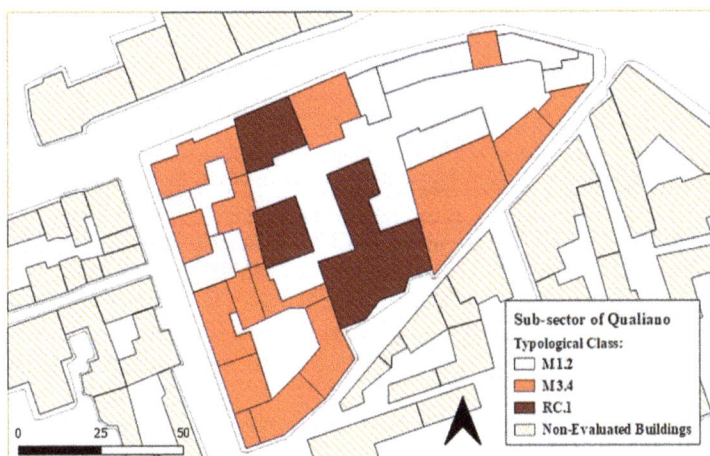

Figure 4. The main typological classes of the examined sub-sector.

2.3. Seismic Vulnerability Assessment

In order to implement a rapid seismic assessment procedure for the examined urban sector the proposed new vulnerability form for aggregates depicted in Table 1 is used [21].

Table 1. The new vulnerability form for building aggregates.

Parameters	Class Score, S_i				Weight, W_i
	A	B	C	D	
1. Organization of vertical structures	0	5	20	45	1.00
2. Nature of vertical structures	0	5	25	45	0.25
3. Location of the building and type of foundation	0	5	25	45	0.75
4. Distribution of plan resisting elements	0	5	25	45	1.50
5. In-plane regularity	0	5	25	45	0.50
6. Vertical regularity	0	5	25	45	0.50–1.00

Table 1. *Cont.*

Parameters	Class Score, S_i				Weight, W_i
	A	**B**	**C**	**D**	
7. Type of floor	0	5	15	45	0.75–1.00
8. Roofing	0	15	25	45	0.75
9. Details	0	0	25	45	0.25
10. Physical conditions	0	5	25	45	1.00
11. Presence of adjacent building with different height	−20	0	15	45	1.00
12. Position of the building in the aggregate	−45	−25	−15	0	1.50
13. Number of staggered floors	0	15	25	45	0.50
14. Structural or typological heterogeneity among adjacent S.U.	−15	−10	0	45	1.20
15. Percentage difference of opening areas among adjacent façades	−20	0	25	45	1.00

This new form is based on the vulnerability index method that was initially proposed by Benedetti and Petrini [6], which was widely used in the past as a rapid technique to investigate the vulnerability of isolated buildings under earthquakes. The basic vulnerability assessment method, composed of ten parameters, was adopted with some minor adjustments by the Italian Group Against Earthquakes as the first screening tool for assessing the vulnerability of masonry and RC buildings.

The first ten parameters of the above mentioned form refer to the main geometrical and mechanical parameters of buildings, as well as to the significant peculiarities of structural systems in the seismic area, such as type and distribution of seismic-resistant elements, foundation category, regularity, floor and roof types, structural detailing, and conservation state.

However, in order to consider the structural interactions among adjacent buildings, which were not considered in the original method, the development of a new form was mandatory. The new investigation form appropriately conceived for building aggregates is obtained by adding to the ten basic parameters of the original form five new parameters taking into account the effects of mutual interaction among the aggregated structural units under seismic actions. The added parameters, partially derived from previous studies found in literature [22], are detailed as follows:

Parameter 11: Presence of adjacent buildings with different height

The in-elevation interaction among adjacent buildings has a not negligible effect on the seismic response of structural units (S.U.) The optimal condition is given by adjoining buildings having the same height (class A), due to the action of mutual confinement. In addition, a building adjacent to higher buildings (from one or both sides, class B) may incur less damage than one adjacent to buildings with less height (classes C and D).

Parameter 12: Position of the building in the aggregate

This parameter aims to take into account the in-plane interaction among S.U. In particular, the isolated building case corresponds to the class D, while the intermediate, corner, and heading conditions are related to the classes A, B, and C, respectively. It is worth noticing that the inclusion in aggregate, independently from the position of the structural unit, gives rise to the seismic vulnerability reduction.

Parameter 13: Number of staggered floors

In case of earthquakes staggered floors are responsible for the pounding effects on masonry walls of adjacent S.U., which can trigger out-of-plane mechanisms. The best situation is the absence of staggered floors (class A), whereas the presence of one (class B), two (class C), or more than two (class D) staggered floors increases the vulnerability.

Parameter 14: Structural or typological heterogeneity among adjacent S.U.

This parameter refers to the possibility that adjacent buildings can be made of different constructive technologies or have structural heterogeneity. In case of buildings of the same masonry type, the vulnerability remains unchanged with respect to the isolated building case (class C). Instead, the case of structural heterogeneity (i.e. a masonry S.U. near to a RC structure) is the most favourable condition in the case of seismic events. Finally, S.U. can be placed close to another unit that is made of masonry stones with worse (class B) or better mechanical properties (class D).

68

Parameter 15: Percentage difference of opening areas among adjacent facades

This parameter influences the seismic response of S.U., since it is responsible for the distribution of horizontal actions among façades of adjacent buildings. Other than the case of no difference of opening areas (class A), it is possible to observe situations where the S.U. is between other buildings with minor (classes B and C) or major (class D) percentage of windows and doors.

Conceptually, the methodology is based on the evaluation of a vulnerability index, I_V, for each S.U. as the weighted sum of the 15 parameters listed in Table 1.

The estimated parameters, are distributed in four classes of increasing vulnerability, (A, B, C, and D), characterised by a score, S_i, and an associated weight, W_i, varying from a minimum of 0.25 (less important parameters), up to a maximum of 1.20 (most important parameters). Further information how scores and classes were determined are found in [7,21,22].

Therefore, the vulnerability index, I_V, can be calculated, as follows:

$$I_V = \sum_{i=1}^{15} S_i \times W_i \tag{1}$$

In order to facilitate its use and interpretation, the vulnerability index value can be normalised in the range (0–1) by means of Equation (2), where it is indicated with the notation V_I.

$$V_I = \left[\frac{I_V - \left(\sum\limits_{i=1}^{15} S_{min} \times W_i \right)}{\left| \sum\limits_{i=1}^{15} \left[(S_{max} \times W_i) - (S_{min} \times W_i) \right] \right|} \right] \tag{2}$$

As illustrated in Figure 5, the application of this procedure to the selected sub-urban sector has allowed to evaluate the seismic vulnerability of masonry and RC buildings located there.

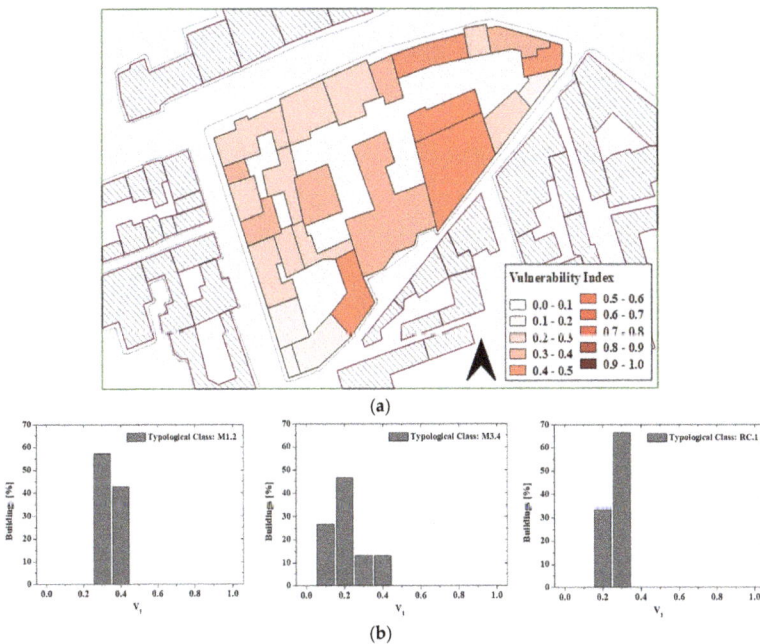

(a)

(b)

Figure 5. Vulnerability analysis results: (**a**) index distribution; (**b**) vulnerability frequencies for each typological classes.

From the application of the vulnerability assessment methodology, it can be seen that, in the study area, the distribution of the vulnerability indexes is heterogeneous with average values of 0.38, 0.26, and 0.30 for typological classes M1.2, M3.4, and RC.1, respectively. The standard deviation (σ_i) associated to the vulnerability index distributions for the analysed typological classes M1.2, M3.4, and RC.1 are respectively $\sigma_{M1.2} = 0.05$ $\sigma_{M3.4} = 0.08$, and $\sigma_{RC.1} = 0.06$.

The synthetic representation of the statistical data takes place while considering the distribution of the total frequencies of the vulnerability indexes that are presented in Figure 5b. It can be noted that 56% of buildings belonging to typological class M1.2 have a vulnerability index of 0.30, while only 44% of them have an index of 0.40. Similarly, for the class M3.4, 45% of buildings have a vulnerability index of 0.20, whereas only a lower percentage of buildings (about 13%) have vulnerability indexes of 0.30 and 0.40. Finally, referring to the RC.1 class, it is possible to notice a vulnerability index of 0.30 for most of the buildings (about 2/3 of the building sample).

2.4. Damage Probability Matrices (DPM) and Vulnerability Curves

The Damage Probability Matrices (DPM) express the occurrence probability of a certain damage level for different typological classes that were subjected to dissimilar seismic intensity levels according to the EMS-98 scale [23]. Methodologically, they can be generated on the basis of a generic damage scale expressed in terms of costs (such as the ratio between the repairing cost and the reconstruction one), which can be understood both in phenomenological terms and according to a qualitative estimate of the different damage degree that buildings may suffer in the case of seismic events [24].

From a practical point of view, the DPM can be implemented after the binomial coefficients are known. In the case under study, DPM of investigated typological classes are plotted in Figure 6.

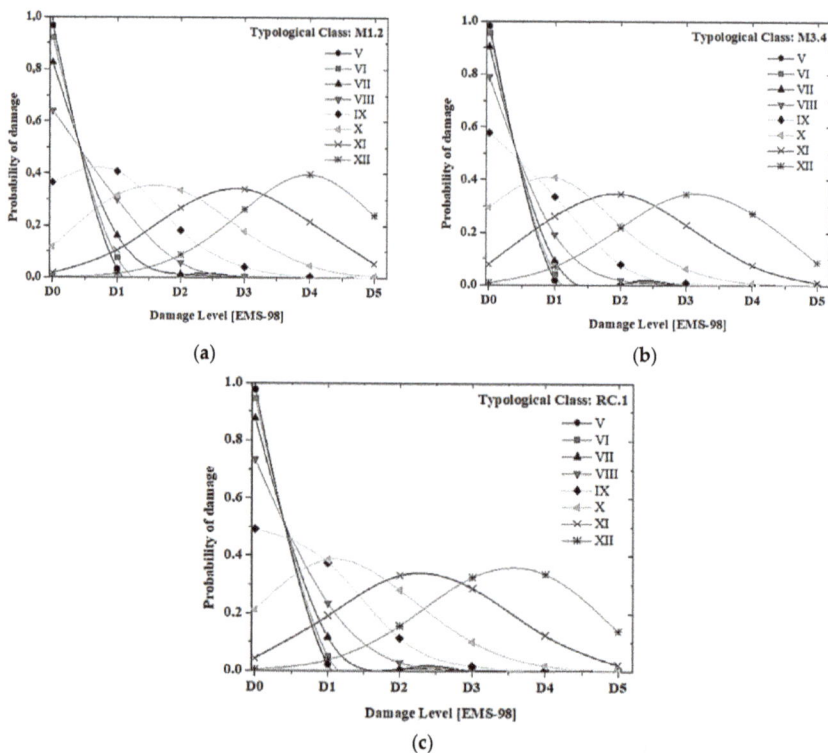

(a)

(b)

(c)

Figure 6. Damage Probability Matrices (DPM) for (a) M1.2, (b) M3.4 and (c) RC.1 typological classes.

In this figure, it is apparent that the binomial distribution has the highest coefficients for the damage degree D0 and intensity $I_{EMS\text{-}98}$ = V. When the seismic intensity increases up to $I_{EMS\text{-}98}$ = XII, the binomial distribution function has a maximum corresponding to the damage grade D4 (partial collapse) for the class type M1.2 with a probability of damage of 40%. Such a probability tends to decrease for the other two typological classes under the same intensity earthquake. It is also important to note that the class type RC.1, if compared to the other typological classes, exhibits an intermediate damage grade D5 (global collapse) due to in-plane and in-elevation irregularities.

The average vulnerability curves [25] are obtained to estimate the propensity of damage of the analysed structural classes. More in detail, these curves express the probability P[SL | $I_{EMS\text{-}98}$] that a building reaches a certain limit state "LS" at a given intensity "$I_{EMS\text{-}98}$" according to the European macroseismic scale (EMS-98).

In particular, as mathematically expressed by Equation (3), the vulnerability curves depend on three variables: the vulnerability index (V_I), the seismic hazard, expressed in terms of macroseismic intensity ($I_{EMS\text{-}98}$), and a ductility factor Q, ranging from 1.0 to 4.0, which describes the ductility of the typological classes. In the case under study, according to [26,27], a ductility factor Q of 2.3 is considered.

$$\mu_D = 2.5 \left[1 + tanh \left(\frac{I_{EMS-98} + 6.25 \times V_I - 13.1}{Q} \right) \right] \tag{3}$$

Recalling the Equation (2), it is also possible to derive vulnerability curves using the mean value and the upper and lower bound ranges of the vulnerability index distribution for different scenarios (V_I-$\sigma_{VI,Mean}$; V_I +$\sigma_{VI,Mean}$; V_I +$2\sigma_{VI,Mean}$; V_I +$2\sigma_{VI,Mean}$). Such a result is presented in Figure 7a–c, respectively, for typological classes M1.2, M3.4, and RC.1.

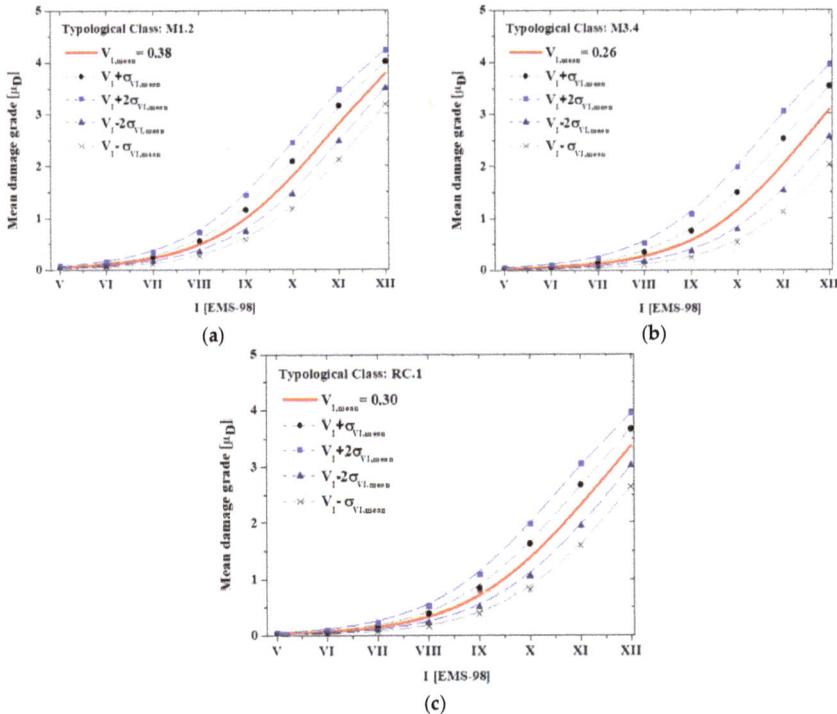

Figure 7. Mean vulnerability curves for building samples of (a) M1.2, (b) M3.4, and (c) RC.1 typological classes.

3. The Damage Scenarios Analysis

The concept of "risk" is generally understood as the economic, social, and environmental consequences that a disastrous event can cause in a given period of time. A prediction of the possible damage scenarios induced by natural events is a useful tool for the quantitative definition of expected losses and for the early implementation of mitigation measures. In this context, the damage analysis allows for an accurate management of the seismic emergency focusing the attention on the urban areas at risk [28]. The severity of the seismic damage is herein analysed by means of a careful parametric analysis. In fact, it should be considered that, during the earthquake, the sample of buildings would be subjected to damages proportional to the seismic motion severity. Using the Gutenberg-Richter law [29], it is possible to predict from the theoretical point of view the possible seismic intensities deriving from magnitudes that can occur in a specific area. To this purpose, the historical earthquakes in the examined area have been taken from the Italian Macroseismic Database DBMI-15 (*National Institute of Geophysics and Volcanology*) [30]. In particular, the seismic events of Piana Campana (1805), Irpinia (1692), and Irpinia-Basilicata (1980), which gave rise to moment magnitudes of 5, 6, and 7, respectively, have been selected (Figure 8). The selection of these magnitude sets has allowed for plotting the expected damage scenarios.

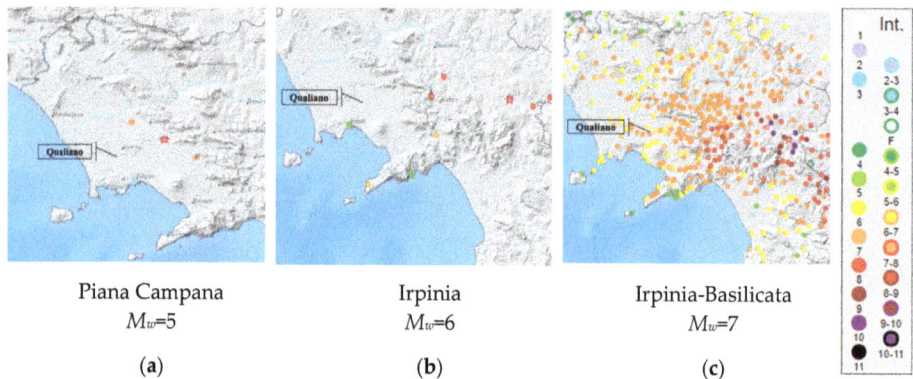

Piana Campana	Irpinia	Irpinia-Basilicata
M_w=5	M_w=6	M_w=7
(a)	(b)	(c)

Figure 8. Historical earthquakes selected for the case study area [30]. (**a**) Piana Campana; (**b**) Irpinia; (**c**) Irpinia-Basilicata.

In Figure 8c, it is worth noting how the Irpinia–Basilicata earthquake generated in the study area a maximum macroseismic intensity, I, equal to VII. Moreover, the municipality of Qualiano is part of the seismogenic zone ZS9-928 (Ischia-Vesuvius) characterised by high values of the expected magnitude. Nevertheless, the cumulative distribution function, F_M (m), for the earthquake magnitudes has been considered according to the following equation:

$$F_M(m) = \frac{1 - 10^{-b(m - m_{min})}}{1 - 10^{-b(m_{max} - m_{min})}}, \quad \forall m_{min} < m < m_{max} \tag{4}$$

where the constant b is generally equal to 1, and m_{min} and m_{max} represent, respectively, the lower and upper limit of the selected magnitude.

The occurrence probabilities of these discrete set of magnitudes, which can be considered as the unique reliable values, are computed using Equation (5) and they are shown in Figure 9.

$$P[M = m_j] = F_M(m_{j+1}) - F_M(m) \tag{5}$$

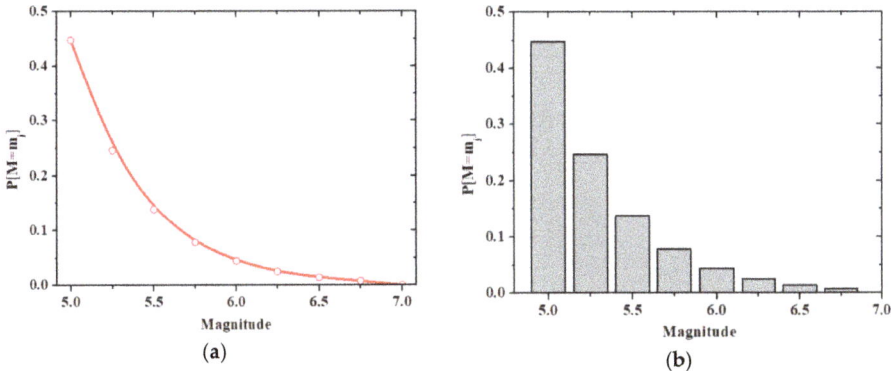

Figure 9. Magnitude distribution for a specific source with a truncated Gutenberg-Richter law:
(a) continuous probability density function; (b) discrete probability density function.

The discrete probability distribution shows that, for a moment magnitude M_W = 5.0, a probability of occurrence of 43% is attained. Instead, the occurrence probability of earthquakes with magnitudes M_W > 5.0 tends to gradually decrease, since these seismic events are very rare.

Aiming at predicting the ground shaking at a site, it is also necessary to model the distribution of distances from the source to the site of interest. For a given earthquake source, it is generally assumed that earthquakes will occur with equal probability at any location on the fault [31].

Given that locations are uniformly distributed, it is generally simple to identify the distribution of source-to-site distances using only the source geometry. The source produces earthquakes randomly and with equal likelihood anywhere within 100 km of the site: the source may be larger, but it is typically truncated at some distance beyond which earthquakes are not expected to cause damage at the site.

From the above considerations, epicentre distances of 5, 10, and 15 km have been considered in order to maximize the expected impact. The attenuation relationship of seismic effects proposed by Esteva [32], as reported in Equation (6), is adopted as follows:

$$I_{EMS-98} = 1.45 \cdot M_w - 2.46 \ln(D) + 8.166 \tag{6}$$

where M_w is the moment magnitude and D is the epicentre distance expressed in kilometres. The correlation obtained between moment magnitude, M_w, and macroseismic intensity, I_{EMS-98}, for different epicentre distances are summarized in Table 2.

Table 2. Correlation between moment magnitude, M_w, and macroseismic intensities, I_{EMS-98}, for different epicentre distances.

Magnitude, M_w	Macroseismic Intensity, I_{EMS-98}		
	D = 5 [Km]	D = 10 [Km]	D = 15 [Km]
5.0	XI	X	IX
6.0	XII	XII	XII
7.0	XII	XII	XII

As it is seen from the data reported in Table 2, when M_w = 5.0 the macroseismic intensity tends to decrease for increasing site-source distances. Contrary, it is worth noting that, for magnitudes 6.0 (detected in 1980 earthquake) and 7.0 (highest grade recorded in the Campania region), the maximum macroseismic intensity, I_{EMS-98} = XII, is reached and it remains as constant independently from the site-source distance considered. Furthermore, referring to the damage parameter μ_D, representative of the damage thresholds of the EMS-98 scale, it is possible to obtain nine damage scenarios, as presented in Figure 10, deriving from the combination of the epicentre distances and the magnitudes considered [33].

D0 (No damage)　D1 (Moderate damage)　D2 (Substantial damage)　D3 (Significant damage)　D4 (Partial collapse)　D5 (Collapse)

D=5 Km; M_w=5.0	D=5 Km; M_w=6.0	D=5 Km; M_w=7.0
D=10 Km; M_w=5.0	D=10 Km; M_w=6.0	D=10Km; M_w=7.0
D=15 Km; M_w=5.0	D=15 Km; M_w=6.0	D=15 Km; M_w=7.0

Figure 10. Damage scenarios of the investigated area within the historical centre of Qualiano.

From the previous figure, it is evident that the damage level does not depend on the site-source distance for magnitudes of 6 and 7, since the occurred macroseismic intensity is always the same. On the other hand, when $M_w = 5$, there is a substantial reduction of damage passing from $D = 5$ to $D = 15$ km. In particular, for an epicentre distance $D = 5$ km and moment magnitude $M_w = 5.0$, it is expected that most of the buildings reach damage thresholds D2 (substantial damage). Instead, when the site-source distance growths up to 10 Km, the most recurrent damage level is D1 (moderate damage). The same result is also attained when $D = 15$ Km, but in this case, contrarily to the previous situation, also buildings without damages (D0 level) are noticed. Differently, with moment magnitude of 6 or 7, independent from the distance value considered, the damage levels D3 (significant damage) and D4 (partial collapse) are expected.

Figure 11 shows the damage distribution obtained for the above described combinations. As it is observed, for $M_w = 5.0$ and $D = 5$ km, it is expected that 50% of the buildings reach damage level D2, 36% of the sample reach damage D3 and only 14% reach damage D1.

For the same epicentre distance, if the magnitude tend to increase either to 6 or 7 most of the buildings, 60% of the sample, reach the stationary damage D3, whereas 40% of buildings attain the damage D4.

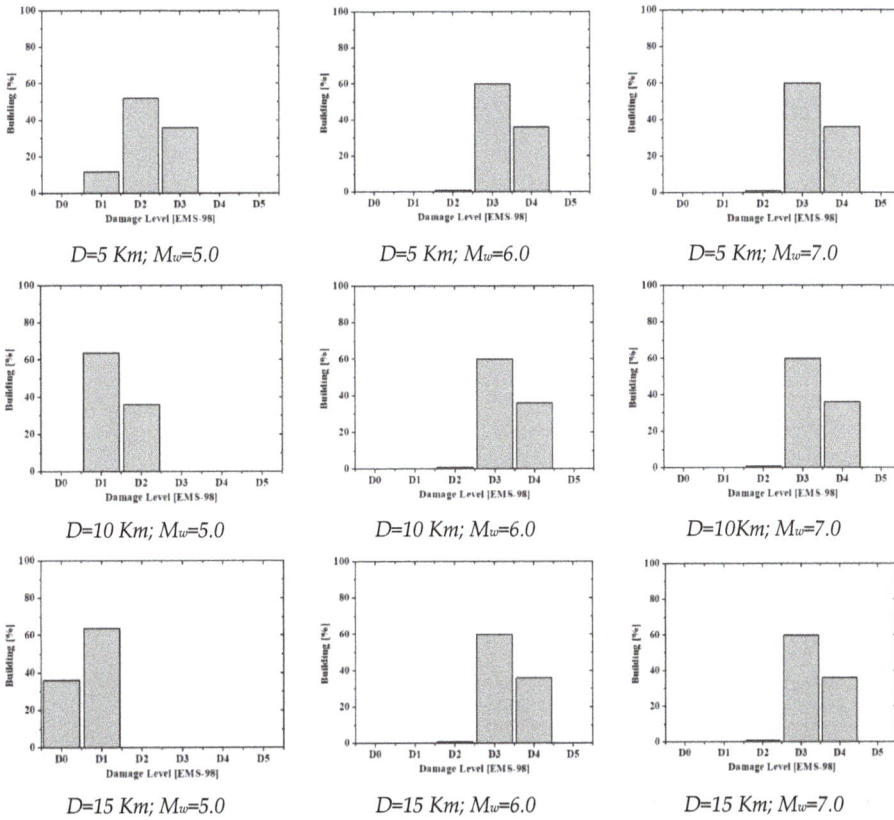

Figure 11. Damage distribution for different epicentre distance and moment magnitude combinations.

4. Geotechnical Hazard Amplification

4.1. The Period-Dependence Site Effect and Local Hazard Amplification

The macroseismic intensity, even though it is a discrete quantity, is the main parameter for correlating the seismic input to the damage deriving from the earthquake impact. Furthermore, as previously analysed, a scenario analysis aims at estimating the global damage level at the territorial scale [34], instead of predicting the response of a given structure at a specific site. However, geotechnical-site characterisation should be considered for more accurate vulnerability and damage distributions. Soil conditions are taken into account in order to incorporate local effects within the vulnerability index method, so that the risk is assessed through macroseismic intensity as compared to rigid soil conditions.

In any case, the study proposed in [35] showed that the macroseismic intensity increase is required by local phenomena, which produce dynamic amplification in terms of vulnerability index modifiers. The macroseismic intensity increment that is induced by geological site phenomena is derived from the period-amplification effects dependence. In particular, referring to a generic elastic spectrum according to the Italian design code [36], a local amplification factor (f_{ag}) can be defined, as follows:

$$f_{ag} = \frac{S_{ae}(T)_K}{S_{ae}(T)_R} \tag{7}$$

75

Being $S_{ae}(T)_K$ the maximum acceleration, S_{ae}, of the elastic spectrum evaluated for a generic class of soil, K, and $S_{ae}(T)_R$ the elastic response spectrum at the bedrock. Therefore, this ratio indicates the local amplification effect due to the generic soil type with respect to the rigid soil condition.

Subsequently, from the knowledge of the amplification factor, f_{ag}, it is possible to estimate the increment of seismic intensity, ΔI, by means of the following equation:

$$\Delta I = \frac{\ln(f_{ag})}{\ln C_2} \tag{8}$$

where the coefficient C_2, estimated equal to 1.82, measures the increase of the seismic acceleration, a_g, with the intensity, I, according to the correlation law proposed in [37].

Besides, the increase of the seismic vulnerability, ΔV (vulnerability modifier), is evaluated by means of the following relationship:

$$\Delta V = \frac{\Delta I}{6.25} \tag{9}$$

Finally, the main vibration period associated to the inspected sample of building is calculated according to the simplified formulation envisaged by [36], as follows:

$$T_1 = \alpha \cdot H^\beta \tag{10}$$

where H is the total height of buildings express in metres, the coefficient α is equal to 0.05 and 0.075 for masonry and RC buildings, respectively, and $\beta = 0.75$ for both building types. Thus, based on these assumptions, the vibration period of masonry buildings is 0.22 seconds, whereas that of RC buildings is 0.48 seconds.

4.2. Evaluation of Site Effect Condition and Seismic Vulnerability Scenarios

The Municipality of Qualiano has an extension of 7.3 Km² and it is located in the centre of the Campana Plain, geologically modified by the intense activity of the Campi Flegrei volcano. Its stratigraphy is mainly composed of celandite deposits mixed with pumiceous lapilli, having black slag, and Campano yellow tuff in the basal area. The geological analyses carried out in [38] using the Multichannel Analysis of Surface Waves (MASW) technique classify the soil as "Category C", that is coarsely thick-grained deposits with thickness exceeding 30 m and shear wave values, V_{S30}, of 341 ms^{-1}, according to the NTC18 [36] (Figure 12).

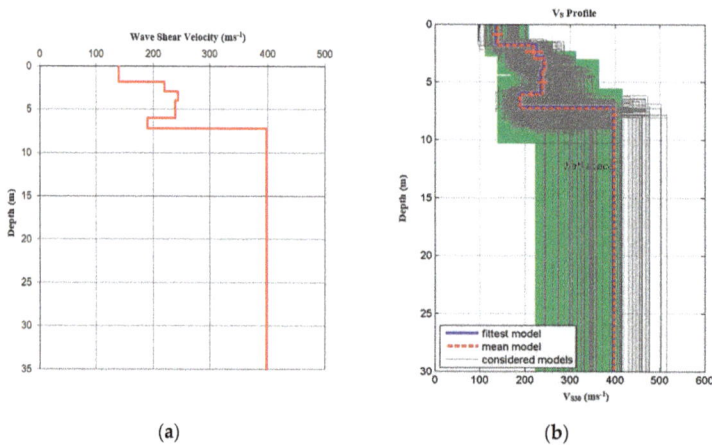

(a)   (b)

Figure 12. Distribution of shear waves: (a) Multichannel Analysis of Surface Waves (MASW) test result; and, (b) median value of shear velocity waves.

The assessment of the vulnerability increase is based on the possible effects induced by vertical and horizontal seismic actions. To this purpose, it is necessary to use, based on the indications that are given in Section 4.1, the local amplification coefficient f_{ag}, as reported in Table 3, where it is noticed that only horizontal accelerations are increased due to local site conditions.

Table 3. Amplification factors due to site effects.

Elastic Spectrum	Sa [T_1]$_C$ [g]	Sa [T_1]$_A$ [g]	f_{ag}
Vertical Spectrum	0.12	0.12	1.0
Horizontal Spectrum	0.52	0.35	1.5

Furthermore, the seismic intensity increase (ΔI), appropriately defined by Equation (8), is used in order to redefine the macroseismic intensity values, previously defined in Section 3, corresponding to the set of magnitudes that were considered at different epicentre distances (Table 4).

Table 4. Macroseismic intensity increase, ΔI, for different scenarios.

Magnitude, M_w	Increase ΔI	Increased Macroseismic Intensity, $I_{EMS-98} + \Delta I$		
		D = 5 [Km]	D = 10 [Km]	D = 15 [Km]
5.0		XII	XI	X
6.0	0.66	XII	XII	XII
7.0		XII	XII	XII

From the results obtained, it is apparent that there is an average increase of 4% of the seismic intensity as compared to the case without site effects.

Analogously, a seismic vulnerability modifier (ΔV_I) can be introduced to take into account the sum of the effects induced by vertical and horizontal seismic action components calculated for soil *category* C. This seismic behaviour modifier is defined through the following equation:

$$\Delta V_I = \sum_i [\Delta V_V + \Delta V_H] \tag{11}$$

Therefore, the global vulnerability of the inspected buildings is calculated as the sum of the local effects contribution and the normalized vulnerability index properly achieved from the building aggregates form:

$$\overline{V_I} = \sum_i [\Delta V_V + \Delta V_H] + V_I = \Delta V_I + V_I \tag{12}$$

The spatial distribution of vulnerabilities with the relative frequencies is indicated in Figure 13.

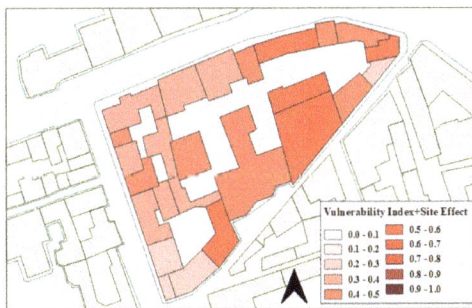

(a)

Figure 13. Cont.

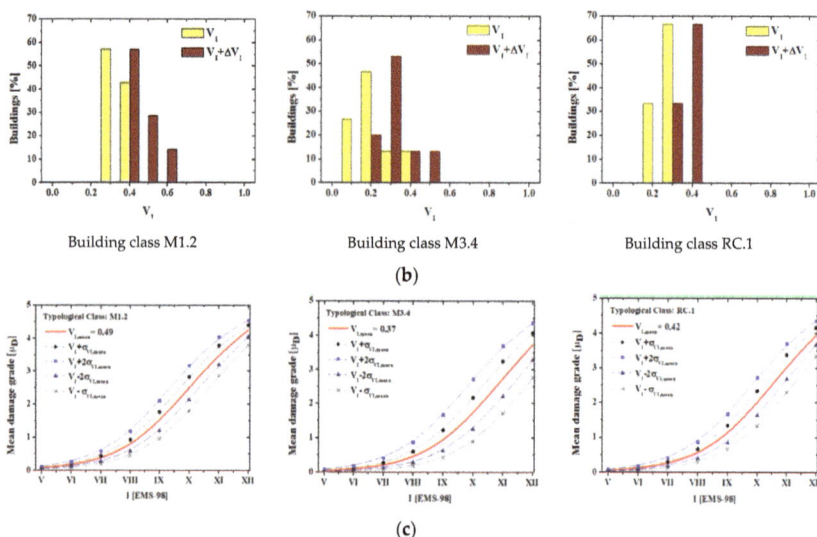

Building class M1.2   Building class M3.4   Building class RC.1

(b)

(c)

Figure 13. Vulnerability distribution in the inspected area (**a**), corresponding vulnerability frequencies (**b**) and mean vulnerability curves of building classes (**c**).

The vulnerability distribution due to site effects (Figure 13a,b) shows a vulnerability increment of 29% for the building class M1.2, of 42% for the M3.4 class and of 36% for the RC.1 class. The standard deviations (σ_i) that are associated to the vulnerability index distributions for the examined building classes are $\sigma_{M1.2} = 0.05$, $\sigma_{M3.4} = 0.08$, and $\sigma_{RC.1} = 0.06$. In Figure 13c, the mean vulnerability curves, assuming the local amplification effects, are shown. In this figure, it is noted that there is an increase of the expected damage, which reaches a level D4 for the class M1.2 and a level included in the range D3÷D4 for both classes M3.4 and RC.1.

Subsequently, in order to predict the damage distribution, the correlation between macroseismic intensity and the seismic acceleration, a_g, i.e. the most used physical parameter of the ground motion, is considered as follows [37]:

$$\log a_g = C_1 \cdot I_{EMS-98} - C_2 [g] \tag{13}$$

where the correlation coefficients, C_1 and C_2, are 0.602 and 7.073, respectively. The used correlation law is graphically shown in Figure 14.

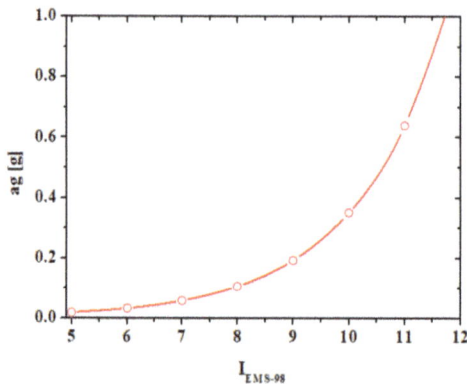

Figure 14. Correlation between macroseismic intensity I_{EMS-98} and a_g.

The non-linear law reported in the previous figure shows that a_g growths in exponential way with the increase of the macroseismic intensity, I_{EMS-98}. In particular, a_g increases from a minimum of 0.025 g for intensity equal to V to a maximum of 1g for intensity slightly less than XII.

The damage distributions, expressed in terms of the expected accelerations for the typological classes analysed, are reported in Figure 15.

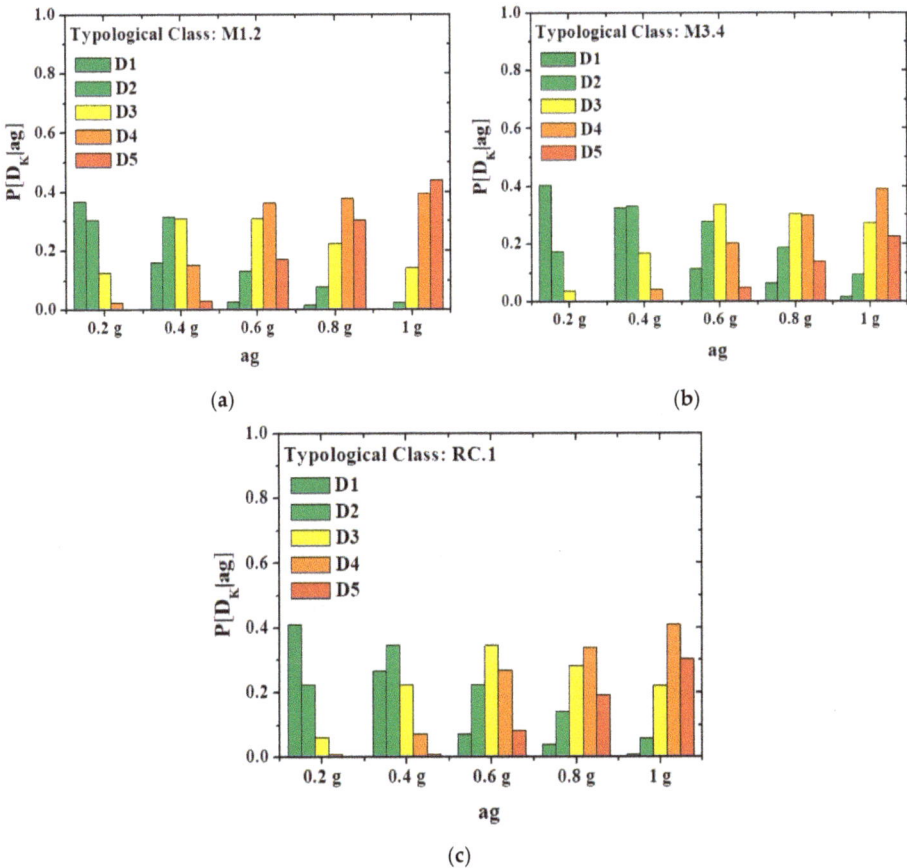

(a)

(b)

(c)

Figure 15. Expected damage distributions for the analysed typological classes at different values of a_g. (a) M1.2; (b) M3.4; (c) RC.1.

The damage distributions show rather heterogeneous results mainly due to different types of buildings examined. It is worth noting that, for a_g equal to 0.2 g, moderate damage condition (D1) is expected with an average occurrence probability of 40% for all building classes. Furthermore, when a_g increases up to a maximum of 1 g, the collapse condition (D5) is expected for the class M1.2 (occurrence probability of 43%), while the partial collapse (D4) is recorded for both typological classes M3.4 and RC.1 (average probability of occurrence of 36%).

5. Conclusions

The study has analysed the seismic vulnerability of a historic sub-urban sector of Qualiano (district of Naples), having a medium to high seismicity with expected PGA values included in the range (0.15g–0.25g), using a parametric-probabilistic approach.

The inspected area is characterised by a structural heterogeneity mainly due to the different constructive techniques that occurred over the years. The characterisation of the building typological classes that are located in the investigated sub-urban sector has been done through the CARTIS form. This form has allowed for grouping buildings into given classes that, due to their construction and manufacturing features, have been considered as homogeneous. Therefore, from the in-situ survey, it has been noticed that in the examined sector buildings are made of tuff masonry structures with different floor types (timber—URM1 class—and reinforced concrete—URM2 class) and reinforced concrete ones (RC class). Subsequently, the URM1, URM2, and RC classes have been named as M1.2, M3.4, and RC.1, according to the acronyms proposed by the Building Typology Matrix of the EMS-98 scale.

Subsequently, the seismic vulnerability of the inspected area has been estimated through a survey form appropriately conceived for building aggregates. The spatial distribution of the vulnerability has been developed by means of the QGIS tool, which provides an average global vulnerability of the examined sub-sector equal to 32%. Subsequently, in order to evaluate the susceptibility at damage of the building sample, the medium vulnerability curves of building classes have been considered. Accordingly, varying the macroseismic intensity of possible earthquakes according to the EMS-98 scale, the expected damages have been estimated for different scenarios using the Gutenberg–Richter law. In particular, by means of this law, it has been possible to determine the discrete distribution of the magnitudes (M_w), selected in the range (5–7), and the relative occurrence probability. In particular, it has been noted that events with moment magnitude greater than 5.0 have a low occurrence probability.

The damage scenarios have been considered using the attenuation law in terms of seismic intensity proposed by Esteva, which proposed a direct correlation between the moment magnitude and the epicentre distance. The results obtained have shown that, for $M_w = 5.0$ the macroseismic intensity tends to decrease when the site-source distance increases. Contrary, for magnitudes 6.0 (detected in 1980 earthquake) and 7.0 (highest grade recorded in the Campania region) the maximum macroseismic intensity, $I_{EMS-98} = XII$, is reached and it remains as constant independently from the site-source distance considered.

On the other hand, with reference to the likely damages in the investigated urban sub-sector, it has been observed that, for moderate values of seismic intensity ($I_{EMS-98} < X$), the probable damages are moderate, but for higher values of seismic intensity ($X < I_{EMS-98} < XII$), significant damages and partial collapses of the analysed building sample should occur. Moreover, the analysis results have shown that M1.2 class buildings are those that are affected by the highest damage level.

Finally, the vulnerability increase induced by site effects has been taken into account in order to properly evaluate the seismic risk of the investigated area for developing possible interventions to reduce the building vulnerability and to safeguard the people live. Therefore, an amplification factor of 1.50 for the horizontal design spectrum has been considered to take into account the site effects of the recorded ground type with reference to the rigid soil condition. The influence of the geological conditions on the vulnerability analysis results has been shown in terms of variation of the expected damage distribution. Thus, it has been detected that local effects provides an increment of 66% of the seismic intensity, while a global vulnerability increment of 36% has been recorded for the investigated typological classes with respect to the case when the site effects are neglected.

Finally, the damage distribution in the building sample under different seismic accelerations has been carried out while taking into account the vulnerability increase due to local amplification effects. From the obtained results, it has been seen that for a_g equal to 0.2 g, moderate damage condition (D1) is expected with an average occurrence probability of 40% for all building classes. Furthermore, when a_g increases up to a maximum of 1 g, the collapse condition (D5) is expected for the class M1.2 (occurrence probability of 43%), while the partial collapse (D4) is recorded for both typological classes M3.4 and RC.1 (average probability of occurrence of 36%).

From the achieved results it has been noticed that site effects play an important role in the vulnerability and risk assessment of urban areas especially under low grade earthquakes.

Neglecting these factors induces considerable inaccuracies in the estimation of the expected seismic damage of buildings of historical centres.

In conclusion, the analysis on the inspected sub-sector, which can be considered as representative of the totality of buildings within the municipality, can be understood as a "pilot research work", providing insights that can be extended to the whole urban centre. However, as future research work, the proposed analysis methodology will be applied to the whole area of the Qualiano's historical centre in order to validate the results on the pilot area.

Author Contributions: Methodology, A.F.; Investigation, N.C.; Data Curation, N.C. and A.F.; Writing-Original Draft Preparation, N.C.; Writing-Review & Editing, A.F.

Funding: This research did not receive any external funding.

Conflicts of Interest: The authors declare no conflict of interest.

References

1. Maio, R.; Ferreira, T.M.; Vicente, R.; Estêvão, J. Seismic vulnerability assessment of historical urban centres: Case study of the old city centre of Faro, Portugal. *J. Risk Res.* **2016**, *19*, 551–580. [CrossRef]
2. Vicente, R.; Parodi, S.; Lagomarsino, S.; Varum, H.; Silva, J.A.R.M. Seismic vulnerability and risk assessment: Case study of the historic city centre of Coimbra, Portugal. *Bull. Earthq. Eng.* **2011**, *9*, 1067–1096. [CrossRef]
3. Kircher, C.A.; Whitman, R.V.; Holmes, W.T. HAZUS Earthquake Loss Estimation Methods. *Nat. Hazards Rev.* **2006**, *7*. [CrossRef]
4. Porter, K.A.; Beck, J.L.; Shaikhutdinov, R. Simplified estimation of economic seismic risk for buildings. *Earthq. Spectra* **2004**, *20*, 1239–1263. [CrossRef]
5. National Group for Protection from Earthquake GNDT. *Manuale per il Rilevamento deLla vulnerabilità Sismica Degli Edifici, Istruzioni per la Compilazione Della Scheda di 2° Livello*; National Group for Protection against Earthquakes GNDT: Roma, Italy, 1993.
6. Benedetti, D.; Petrini, V. Sulla vulnerabilità si sismica di edifici in muratura: Un metodo di valutazione. *L'Industria delle Costr.* **1984**, *149*, 64–72.
7. Formisano, A.; Florio, G.; Landolfo, R.; Mazzolani, F.M. Numerical calibration of an easy method for seismic behaviour assessment on large scale of masonry building aggregates. *Adv. Eng. Softw.* **2015**, *80*, 116–138. [CrossRef]
8. Ferreira, T.M.; Maio, R.; Vicente, R. Analysis of the impact of large scale seismic retrofitting strategies through the application of a vulnerability-based approach on traditional masonry buildings. *Earthq. Eng. Eng. Vib.* **2017**, *16*, 329–348. [CrossRef]
9. Formisano, A.; Chieffo, N.; Mosoarca, M. Seismic vulnerability and damage speedy estimation of an urban sector within the municipality of San Potito Sannitico (Caserta, Italy). *Open Civ. Eng. J.* **2017**, *17*, 1106–1121. [CrossRef]
10. Formisano, A. Theoretical and Numerical Seismic Analysis of Masonry Building Aggregates: Case Studies in San Pio Delle Camere (L'Aquila, Italy). *J. Earthq. Eng.* **2017**, *21*, 227–245. [CrossRef]
11. Giovinazzi, S.; Lagomarsino, S. A Method for the Vulnerability Analysis of Built-up areas. In Proceedings of the International Conference on Earthquake Loss Estimation and Risk Reduction, Bucharest, Romania, 24–26 October 2002.
12. Lagomarsino, S.; Giovinazzi, S. Macroseismic and mechanical models for the vulnerability and damage assessment of current buildings. *Bull. Earthq. Eng.* **2006**, *4*, 415–443. [CrossRef]
13. Faccioli, E.; Pessina, V.; Calvi, G.M.; Borzi, B. A study on damage scenarios for residential buildings in Catania city. *J. Seismol.* **1999**, *3*, 327–343. [CrossRef]
14. Ambraseys, N.; Douglas, J. Near-field horizontal and vertical earthquake ground motions. *Soil Dyn. Earthq. Eng.* **2003**, *23*, 1–18. [CrossRef]
15. Giovinazzi, S.; Balbi, A.; Lagomarsino, S. Un modello di vulnerabilità per gli edifici nei centri storici. In Proceedings of the XI ANIDIS, Congr. Naz. Ingegneria Sismica Ital., Genova, Italy, 25–29 January 2004.
16. Sabatino, G. *Ipotesi Storico-Urbanistiche Sull'origine e Sullo Sviluppo Della Città di Qualiano*; Istituto di Studi Atellani: Casoria, Naples, Italy, 1986.

17. Zuccaro, G.; Della Bella, M.; Papa, F. Caratterizzazione tipologico strutturali a scala nazionale. In Proceedings of the 9th National Conference ANIDIS, L'ingegneria Sismica in Italia, Torino, Italy, 20–23 September 1999.

18. Cacace, F.; Zuccaro, G.; De Gregorio, D.; Perelli, F.L. Building Inventory at National scale by evaluation of seismic vulnerability classes distribution based on Census data analysis: BINC procedure. *Int. J. Disaster Risk Reduct.* **2018**, *28*, 384–393. [CrossRef]

19. Mouroux, P.; Le Brun, B. Presentation of RISK-UE project. *Bull. Earthq. Eng.* **2006**, *4*, 323–339. [CrossRef]

20. QGIS Development Team. QGIS Geographic Information System. Open Source Geospatial Foundation Project. 2014. Available online: http://qgis.osgeo.org (accessed on 26 November 2018).

21. Formisano, A.; Mazzolani, F.M.; Florio, G.; Landolfo, R. A quick methodology for seismic vulnerability assessment of historical masonry aggregates. In Proceedings of the COST Action C26 Urban Habitat Constr. under Catastrophic Events, Naples, Italy, 16–18 September 2010.

22. Formisano, A.; Florio, G.; Landolfo, R.; Mazzolani, F.M. Numerical calibration of a simplified procedure for the seismic behaviour assessment of masonry building aggregates. In Proceedings of the 13th International Conference on Civil, Structural and Environmental Engineering Computing, Chania, Crete, Greece, 6–9 September 2011.

23. Grünthal, G. (Ed.) *Chaiers du Centre Européen de Géodynamique et de Séismologie: Volume 15—European Macroseismic Scale 1998*; European Center for Geodynamics and Seismology: Luxembourg, 1998; ISBN 2879770084.

24. Goretti, A. The Italian contribution to the USGS PAGER project. In Proceedings of the 14th World Conference on Earthquake Engineering, Beijing, China, 12–17 October 2008.

25. Vicente, R.; Ferreira, T.M.; Maio, R. Seismic Risk at the Urban Scale: Assessment, Mapping and Planning. *Procedia Econ. Financ.* **2014**, *18*, 71–80. [CrossRef]

26. Giovinazzi, S.; Lagomarsino, S.; Pampanin, S. Vulnerability Methods and Damage Scenario for Seismic Risk Analysis as Support to Retrofit Strategies: An European Perspective. In Proceedings of the NZSEE Conference, Napier, New Zealand, 8–10 March 2006.

27. Lagomarsino, S. On the vulnerability assessment of monumental buildings. *Bull. Earthq. Eng.* **2006**, *4*, 445–463. [CrossRef]

28. Uva, G.; Sanjust, C.A.; Casolo, S.; Mezzina, M. ANTAEUS Project for the Regional Vulnerability Assessment of the Current Building Stock in Historical Centers. *Int. J. Archit. Herit.* **2016**, *10*. [CrossRef]

29. Wesnousky, S.G. The Gutenberg-Richter or characteristic earthquake distribution, which is it? *Bull. Seismol. Soc. Am.* **1994**, *84*, 1940–1959.

30. National Institute of Geophysics and Volcanology (INGV). Macroseismic Italian Database DBMI-15 release 1.5. 2015. Available online: emidius.mi.ingv.it/CPTI15-DBMI15/ (accessed on 26 November 2018). (In Italian)

31. Lin, T.; Baker, J. Probabilistic seismic hazard deaggregation of ground motion prediction models. In Proceedings of the 5th International Conference Earthquake Geotechnical Engineering, Santiago, Chile, 10–13 January 2011.

32. Jaimes, M.A.; Reinoso, E.; Esteva, L. Risk analysis for structures exposed to several multi-hazard sources. *J. Earthq. Eng.* **2015**, *19*, 297–312. [CrossRef]

33. Formisano, A.; Chieffo, N. Expected seismic risk in a district of the Sant'Antimo's historical centre. *Trends Civ. Eng. its Archit.* **2018**, *2*, 1–15.

34. Formisano, A. Local- and global-scale seismic analyses of historical masonry compounds in San Pio delle Camere (L'Aquila, Italy). *Nat. Hazards* **2017**, *86*, 465–487. [CrossRef]

35. Giovinazzi, S. Geotechnical hazard representation for seismic risk analysis. *Bull. N. Zeal. Soc. Earthq. Eng.* **2009**, *42*, 221–234.

36. Ministerial Decree, D.M. Updating of Technical Standards for Construction. 2018. Available online: http://www.gazzettaufficiale.it/eli/gu/2018/02/20/42/so/8/sg/pdf (accessed on 27 September 2018).

37. Guagenti, E.; Petrini, V. The Case of Old Buildings: Towards a New Law—Intensity Damage. In Proceedings of the 12th Italian Conference on Earthquake Engineering, ANIDIS, Italian National Association of Earthquake Engineering, Pisa, Italy, 10–14 June 1989. (In Italian)
38. Piano Urbanistico Comunale (P.U.C). 2018. Available online: http://www.comune.qualiano.na.it (accessed on 26 November 2018).

buildings

MDPI

Article

Feasibility of Using Neural Networks to Obtain Simplified Capacity Curves for Seismic Assessment

João M. C. Estêvão

DEC-ISE, University of Algarve, 8005-139 Faro, Portugal; jestevao@ualg.pt; Tel.: +351-289-800-154

Received: 21 September 2018; Accepted: 2 November 2018; Published: 6 November 2018

Abstract: The selection of a given method for the seismic vulnerability assessment of buildings is mostly dependent on the scale of the analysis. Results obtained in large-scale studies are usually less accurate than the ones obtained in small-scale studies. In this paper a study about the feasibility of using Artificial Neural Networks (ANNs) to carry out fast and accurate large-scale seismic vulnerability studies has been presented. In the proposed approach, an ANN was used to obtain a simplified capacity curve of a building typology, in order to use the N2 method to assess the structural seismic behaviour, as presented in the Annex B of the Eurocode 8. Aiming to study the accuracy of the proposed approach, two ANNs with equal architectures were trained with a different number of vectors, trying to evaluate the ANN capacity to achieve good results in domains of the problem which are not well represented by the training vectors. The case study presented in this work allowed the conclusion that the ANN precision is very dependent on the amount of data used to train the ANN and demonstrated that it is possible to use ANN to obtain simplified capacity curves for seismic assessment purposes with high precision.

Keywords: vulnerability assessment; capacity curves; neural networks; earthquakes

1. Introduction

There are many methods used for the seismic vulnerability assessment of buildings. These methods can be classified as empirical, analytical/mechanical or hybrid methods [1] and present different levels of accuracy, being the analytical/mechanical methods the most accurate and the empirical the less ones. The selection of a given method is mostly dependent on the scale of the analysis and on the knowledge level about the building's characteristics. If the study is carried out at a building scale (just one single structure) and there is a full knowledge about the geometry, details and materials, it is possible to carry out a very precise seismic analysis, namely using a nonlinear method, which obviously will increase the reliability of the obtained results. However, at urban scale (one city or part of it), or at regional scale (a set of cities), or even at global scale (a country or even a continent), it is almost impossible to adopt the same detailed approach, so it is frequent to adopt empirical methods which are based on the damage observed after earthquakes, which normally exhibit a high dispersion level. It is obvious that the volume of data that is necessary to collect and the computational effort necessary to process the amount of data is not equal for all the aforementioned scales. This fact implies that the results obtained in large-scale studies are normally less accurate than those obtained in small-scale studies. For this reason, buildings are usually categorized in several different typologies, for example depending on the age of the construction or on the building materials and structural system and a mean seismic structural behaviour is usually considered for the buildings of a given typology.

A considerable amount of research has been carried out in the last decades on trying to improve the precision of the seismic vulnerability assessment results for large-scale studies. There are many computer programs developed to carry out seismic risk analysis using empirical and/or analytical/mechanical methods, such as the software HAZUS-MH [2] or the software ELER [3], which are examples of computer programs developed to allow to perform large-scale studies, namely supporting different levels of accuracy.

There are also other hybrid approaches to the problem and some of them have adopted methods and techniques of the so-called Artificial Intelligence (AI), such as Artificial Neural Networks (ANNs), which are computational approaches that try to imitate the brain.

Nowadays ANNs have been used for solving a variety of problems, such as visual recognition, speech recognition and natural language processing, so general public is familiar to the capabilities of modern ANNs in their daily life, even though not understanding how it works, because modern cars may also use ANNs in software for traffic sign recognition [4].

For these reasons, modern ANN capabilities may be an important help in solving many complex problems of earthquake engineering, which feasibility should be investigated.

The use of ANNs applied to civil engineering problems is not a new subject [5], nor their use in the development of computer software for seismic risk assessment [6], using these less traditional techniques within the scope of earthquake engineering, with the aim of trying to improve the reliability of the results.

In the last decades, several studies have been published for structural behaviour evaluation using neural networks, namely for earthquake engineering applications. ANNs have been used to predict the linear [7] and the nonlinear [8,9] dynamic responses of structures subject to earthquakes for damage assessment [10–13], namely using fragility curves [14,15], or for seismic reliability assessment [16,17]. A Monte Carlo simulation technique was also adopted for generating data used for training ANNs [18].

Traditional seismic vulnerability assessment methods that use a mean capacity curve (which is representative of a given structural typology) for estimating seismic vulnerability and earthquake damage have some problems in the structural performance evaluation of an individual building. The main reason is related to the dispersion of values around the mean curve. The real capacity curve of a given building will probably be different from the typological mean capacity curve. This means that the average result of a typical typological capacity curve can lead to overestimating or underestimating the real seismic damage, depending on the studied building.

In this work, a study about the feasibility of using Multi-layer Feed-Forward Neural Network (MFFNN) to obtain a simplified capacity curve of a given building typology is presented, trying to reduce the results dispersion normally associated to the use of a mean typological capacity curve. The ANNs were previously trained with the results of the nonlinear analysis carried out for several structures of a given typology. A sensitivity analysis was carried out to understand the effect of the number of training vectors in the results precision by comparing the results of an analysis of variance (ANOVA) applied to the outputs of two different ANNs.

Seismic structural performance point (the interception between the demand nonlinear response spectrum and the capacity curve of the structure) for a given seismic action can be obtained using the N2 method [19]. This is a very simple and fast method for seismic nonlinear static (pushover) global analysis, that is presented in the Annex B of the Eurocode 8 (EC8) [20] to find the target displacement d_t (corresponding to the EC8 performance point). The results obtained in this work indicate that the combination of the N2 method with a simplified capacity curve obtained from an ANN allows fast structural seismic performance evaluation on large-scale studies, with an accuracy level of the results much closer to the obtained in a study carried out at a building scale (by minimizing the dispersion values) than the ones obtained with a mean typological capacity curve.

The obtained results open the path for the development of more complex ANN architectures and considering much more input variables.

2. Simplified Capacity Curves and the N2 Method

According to the N2 method, the first step to compute the performance point of a given structure is to obtain a simplified equivalent elastic-perfectly plastic capacity curve of a one degree of freedom dynamic system. In this work, the simplified capacity curve was obtained so that the linear branch corresponds to the one associated to the maximum base shear force (Figure 1).

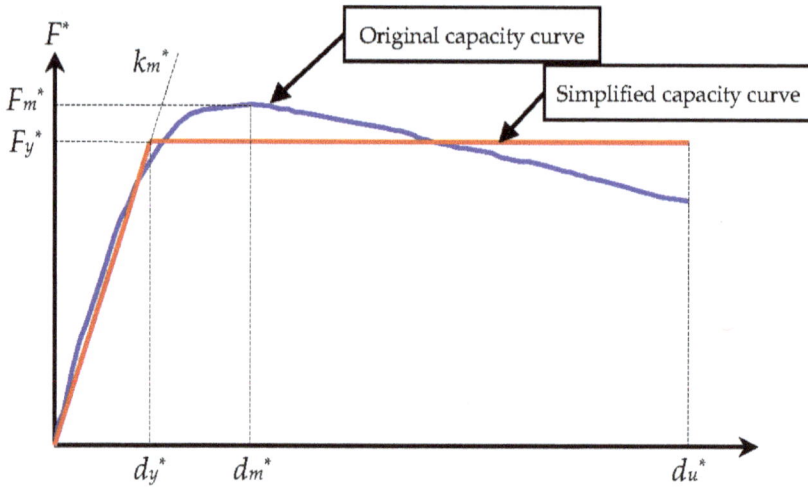

Figure 1. Original and simplified adopted capacity curve.

The stiffness $k_m{}^*$, force $F_y{}^*$ and the displacement $d_y{}^*$, as presented in Figure 1, can be determined throughout the following equations:

$$k_m^* = \frac{F_m^*}{2 \cdot \left(d_m^* - \frac{E_m^*}{F_m^*} \right)},$$

(1)

$$F_y^* = k_m^* \cdot \left[d_u^* - \sqrt{\frac{k_m^* \cdot (d_u^*)^2 - 2 \cdot E_u^*}{k_m^*}} \right],$$

(2)

$$d_y^* = \frac{F_y^*}{k_m^*},$$

(3)

being $E_u{}^*$ the total area limited by the original capacity curve until the ultimate displacement $d_u{}^*$ (which is equal to the total area limited by the simplified capacity curve) and $E_m{}^*$ the area limited by the original capacity curve until the displacement $d_m{}^*$.

After the determination of the simplified capacity curve, the performance point can be obtained using the procedure presented in the Annex B of the EC8 [20].

3. Artificial Neural Networks and Capacity Curves

Software development for fast structural vulnerability assessment is very important for civil protection purposes. This type of software allows mapping the buildings where damage is likely to occur. The computer strategy proposed in this study is to use the mapping capabilities of the ANN to improve the speed and the accuracy of the seismic vulnerability assessment of many buildings. The idea is to use several easy to measure building characteristics to obtain the simplified building capacity curve of a given typology.

ANNs are techniques inspired in biological systems, which have generalization capabilities, so they can be used in structural analysis problems that obey to certain rules (that can be unknown), which are learned during the ANN training process [21].

There are many different types of ANNs. In this work, a general MFFNN is proposed to obtain the parameters of a simplified capacity curve (Figure 2), which is previously trained with the error back-propagation algorithm.

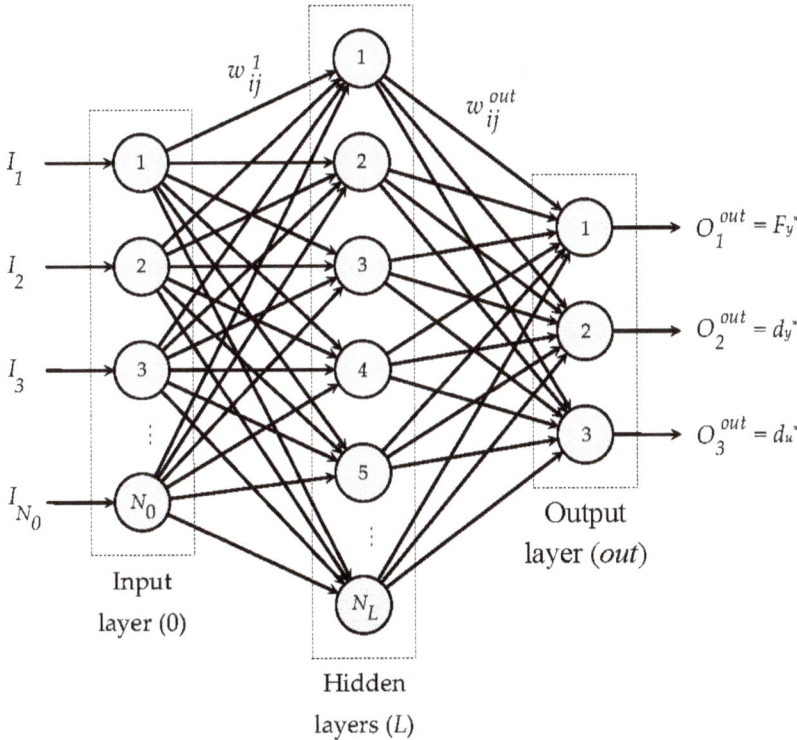

Figure 2. General proposal of a MFFNN used to obtain the building simplified capacity curve.

Previous studies have shown that this type of ANN has good capabilities to capture the structural nonlinear behaviour [22].

The artificial neuron (AN) of an ANN is a computational element which transforms the input signals (x_i) in an output result (like a brain cell). Each AN has an activity v_j (Figure 3), which is equal to

$$v_j = w_{0j} + \sum_{i=1}^{N} x_i \cdot w_{ij}. \qquad (4)$$

The output result depends on the computation of the activation function $f(v)$. A sigmoid function was adopted in this study:

$$f(v) = \frac{1}{1 + e^{-v}}. \qquad (5)$$

The ANN training algorithm implemented in the NEUNET computer program [23], that was used in this work, involves the following steps:

- The input variables values (I_j) and the known output results (the training vectors T_j, which are the simplified capacity curve parameters) are normalized (between 0 and 1) to the maximum values;
- The initial weights are set to a random small value (between −0.5 and +0.5);
- The input values (I_j) of the N_{trn} training vectors are set at the N_0 neurons of the input layer and the known results T_j of these vectors are set at the output layers;

- For each training vector ($n = 1, \ldots, N_{trn}$) and for each neuron j of the layer $L = 1$, the neuron activity and the output are determined, being

$$v_{j,n}^1 = w_{0j}^1 + \sum_{i=1}^{N_0} I_{i,n} \cdot w_{ij}^1, \tag{6}$$

$$O_{j,n}^1 = f\left(v_{j,n}^1\right), \tag{7}$$

and for the other levels ($L = 2, \ldots, out$)

$$v_{j,n}^L = w_{0j}^L + \sum_{i=1}^{N_{L-1}} O_{i,n}^{L-1} \cdot w_{ij}^L, \tag{8}$$

$$O_{j,n}^L = f\left(v_{j,n}^L\right); \tag{9}$$

- Weights are corrected based on the known output results and on the following error expressions, beginning from the output layer

$$D_{i,n}^{out} = \left(T_{i,n} - O_{i,n}^{out}\right) \cdot \left(1 - O_{i,n}^{out}\right) \cdot O_{i,n}^{out}, \tag{10}$$

and following by the other existing levels ($L = 1, \ldots, out - 1$)

$$D_{i,n}^L = \left(\sum_{j=1}^{N_{L+1}} D_{j,n}^{L+1} \cdot w_{ij}^{L+1}\right) \cdot \left(1 - O_{i,n}^L\right) \cdot O_{i,n}^L, \tag{11}$$

$$Z_{ij,n}^1 = D_{j,n}^1 \cdot I_{i,n}, \tag{12}$$

$$Z_{ij,n}^L = D_{j,n}^L \cdot O_{i,n}^{L-1}, \tag{13}$$

- Finely, the new weights are determined

$$new\_w_{ij}^L = actual\_w_{ij}^L + \eta \cdot \sum_{n=1}^{N_{trn}} Z_{ij,n}^L + \alpha \left(actual\_w_{ij}^L - old\_w_{ij}^L\right). \tag{14}$$

The learning parameter η rules the algorithm convergence rate. This rate is lower when using very small η values and it increases with higher η values. A momentum factor ($0 \leq \alpha \leq 1$) can also be used to increase the algorithm convergence rate.

The ANN error obtained for each training vector is equal to

$$E_n = \frac{1}{2} \cdot \sum_{L=1}^{out} \sum_{j=1}^{N_L} \left(D_{j,n}^L\right)^2, \tag{15}$$

and the ANN global error is equal to

$$E = \sum_{n=1}^{N_{trn}} E_n. \tag{16}$$

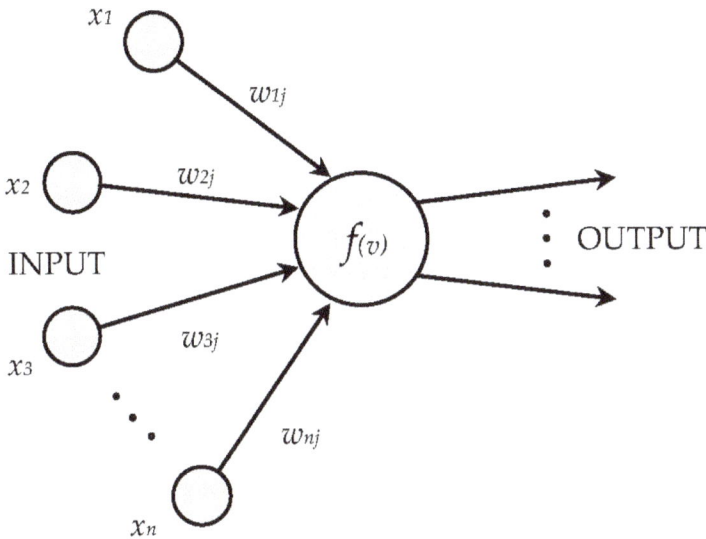

Figure 3. Single artificial neuron representation.

A correct ANN level of training is very important to assure a good reproduction of the mean simplified capacity spectrum of a given structure. A poor ANN training will lead to unsatisfactory results. However, the consequences of an over-training of the ANN can lead to even worst results, because the ANN will adjust the simplified capacity spectrum curve to the local values instead to the mean values of the given typology. In Figure 4, the left ANN solution is an example of a good result and the right ANN solution is an example of a bad result, because the error is too high.

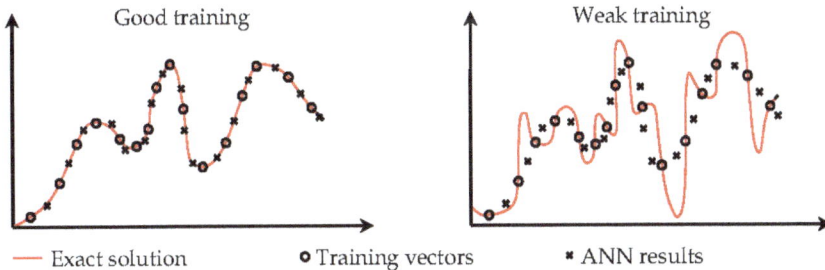

Figure 4. Example of the influence of the ANN training in the output results.

The training problem exemplified in Figure 4 is much more important when the whole region of possible solutions is not well covered by the training vectors, because there are many possible solutions that fit those points. In spite of the evidence that an appropriated training of the ANN can lead to better results than some traditional approaches for seismic assessment of large number of buildings [12], even when considering a relatively small number of training vectors, it is important to understand how feasible the use of an ANN for practical problems with different number of training vectors is.

4. Case Study

Damage dispersion observed in buildings that were affected by earthquakes in the past can be related to seismic vibrations characteristics (influenced by source characteristics and by geological site conditions) and to the differences in the seismic vulnerability of each building [24]. For this reason, it is very important to do an adequate assessment of this vulnerability, namely by using nonlinear seismic analysis, being very common the use of an incremental nonlinear analysis [25] or a static (pushover) nonlinear analysis [26] for vulnerability assessment purposes. When assessing an individual building and when the knowledge level is low, the part 3 of Eurocode 8 (EC8-3) proposes an assessment based on simulated design in accordance with usual practice at the time of construction, which was the strategy adopted in this case study but for a more general purpose.

The present case study aims to evaluate how feasible is to use an ANN (trained with a low number of training vectors) for vulnerability assessment, in terms of accuracy of results.

4.1. Studied Structural Typology

Probably, the concrete buildings with higher seismic vulnerability are the ones built prior to modern seismic codes. Between the decades of 1930s and 1950s, in countries like Portugal, the buildings were designed without considering any seismic action and that is the reason why the proposed approach was tested in this specific typology.

As it has been possible to observe in old structural designs and according to the codes of that period [27], the area of the reinforcing steel bars (A_s) was determined considering an equivalent concrete area of the homogenized cross-section (usually using a homogenization factor of 10 for beams and 15 for columns).

Due to the lack of computational resources, the axial forces (N_c) were normally determined by multiplying the influence area of each column by the weight of the floor. The compression stresses were determined assuming an elastic behaviour of the homogenized cross-section. So, these simplified assumptions led to the necessity of a very small amount of reinforcement in low-rise buildings and normally the value of A_s was just a minimum percentage of the concrete cross-section, which creates very vulnerable buildings, in terms of their seismic behaviour.

In this work, this old simplified procedure was used to design some reinforced concrete frames (Figure 5) in order to simulate the design solutions usually adopted in Portugal in that period (using the minimum number of rebars that leads to $A_s \geq 0.005 \cdot b_c \cdot h_c$), which were used as the training set of a MFFNN. The adopted input variables were the number of beam spans ($I_1 = n_b$), the mean beam span dimension ($I_2 = L_b$) and the mean cross-section column height ($I_3 = h_c$).

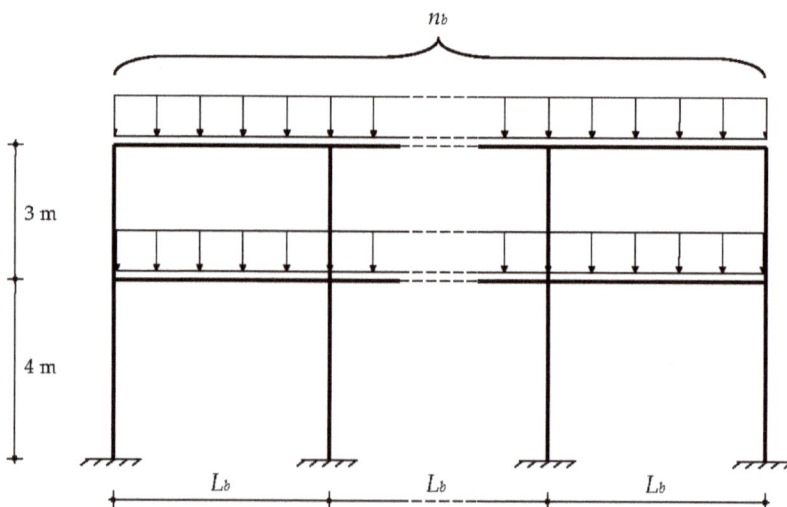

Figure 5. Schematization of the studied concrete frames.

A design value of 40 kgf/cm² was used for concrete in pure compression, a value of 45 kgf/cm² was used in flexure and a design value of 1200 kgf/cm² was used for the reinforcement [27].

The slabs thickness and the beams high were considered as a function of L_b, so the mass per unit area was also considered as a function of L_b. As a simplification and just for the purpose of this study, a constant value of 0.25 m was adopted for the width of all beams (b_b) and columns (b_c). A T-section was adopted for all concrete beams, as proposed in the EC8.

The mass adopted for each dynamic structural system was computed considering a transversal influence area equal to L_b.

At first, 125 nonlinear static analyses were carried out, which were named as the training set n. 1 (TS1), considering frames with the following values: n_b = 1, 2, 3, 4 and 5 spans; L_b = 2, 3, 4, 5 and 6 m; h_c = 0.25, 0.325, 0.4, 0.475 and 0.55 m.

The capacity curves were obtained by using the SeismoStruct software [28] and adopting a triangular force pattern. In Figure 6a all the 125 original capacity curves of the single degree of freedom system are presented and the corresponding simplified equivalent elastic-perfectly plastic capacity curves are presented in Figure 6b.

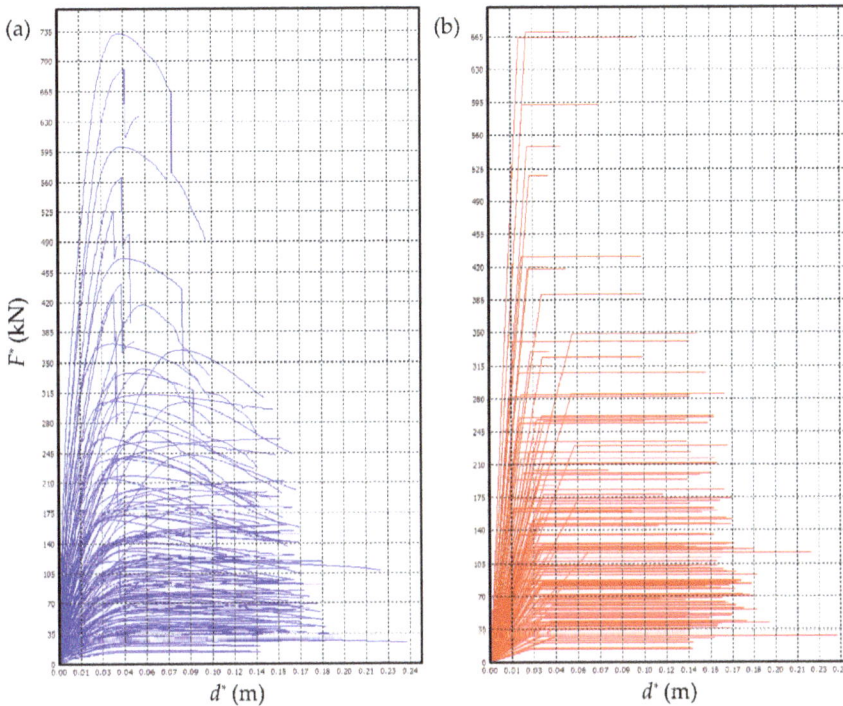

Figure 6. Capacity curves obtained for the (**a**) original structural system; and (**b**) for the simplified equivalent elastic-perfectly plastic system.

Each incremental static nonlinear structural analysis was carried out until the EC8-3 near collapse (NC) limit state was reached for the chord rotation capacity (Equation (A.1) of the EC8-3), or when it was impossible to reach the convergence of iterative process used in the nonlinear structural analysis. When the NC shear capacity limit was reached (Equation (A.12) of EC8-3), the shear strength was reduced to a value corresponding to only 20% of the original strength (this is the SeismoStruct default option, which seems acceptable when observing some laboratorial tests results [29]).

Another training set with 27 capacity curves was also considered, which was named as the training set n. 2 (TS2), which was a subset of the first one, with: n_b = 1, 3 and 5 spans; L_b = 2, 4 and 6 m; h_c = 0.25, 0.4 and 0.55 m.

Additionally, three control cases (not belonging to any of the training sets TS1 or TS2) were considered: the control case n. 1 (CC1), with n_b = 1, L_b = 5.5 m and h_c = 0.5 m; the control case n. 2 (CC2), with n_b = 3, L_b = 4.2 m and h_c = 0.35 m; and the control case n. 3 (CC3), with n_b = 4, L_b = 3.5 m and h_c = 0.28 m.

To process such an amount of data, computer procedures were developed for the automatic creation of computer files containing all the training set values.

4.2. Adopted MFFNN

The training of a MFFNN is not an exact science, in spite of the many approaches available to optimize this process [30] and also depends on the experience obtained in past studies. In this work, a trial and error process were adopted to minimize the ANN output error. The neural network adopted in this study is presented in Figure 7.

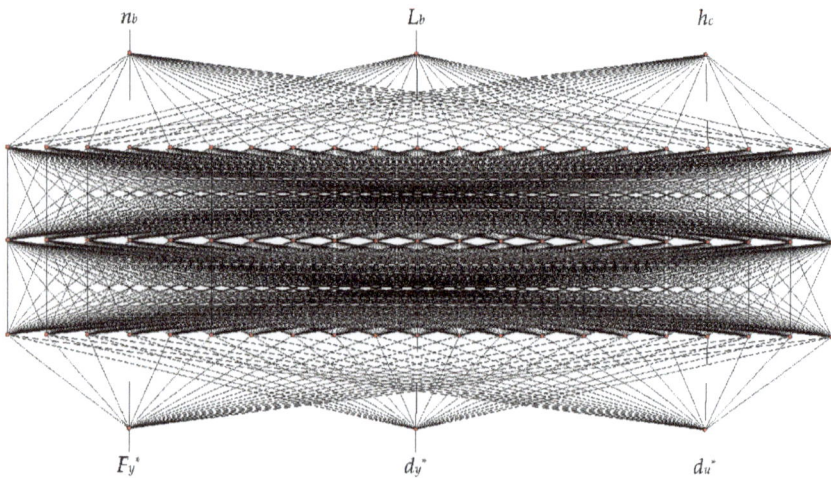

Figure 7. Adopted neural network architecture.

The selection of the number of hidden layers and neurons was carried out by increasing those numbers until a compromise between precision and training speed was reached.

Two artificial neural networks (ANN1 and ANN2) with the same architecture (Figure 7) were trained for this study, using the NEUNET computer program [23]. The first (ANN1), was trained using all the 125 training vectors (TS1) and the other (ANN2) was trained using only a subset of 27 training vectors (TS2), as earlier described.

4.3. Results and Discussion

The results obtained with the two neural networks (ANN1 and ANN2) are compared in Figures 8–10. The blue dots are the results obtained from the 125 nonlinear structural analysis (TS1) and the red and green lines are the corresponding results obtained with ANN1 and ANN2, respectively.

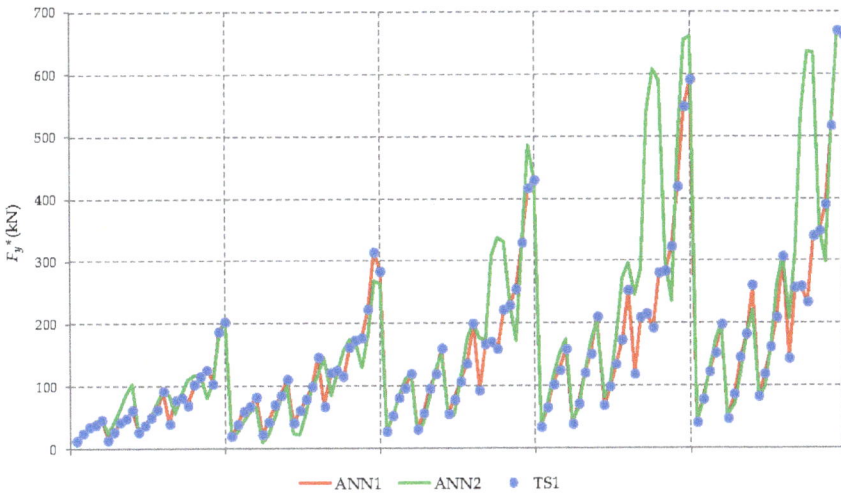

Figure 8. Results comparison for the $F_y{}^*$ values.

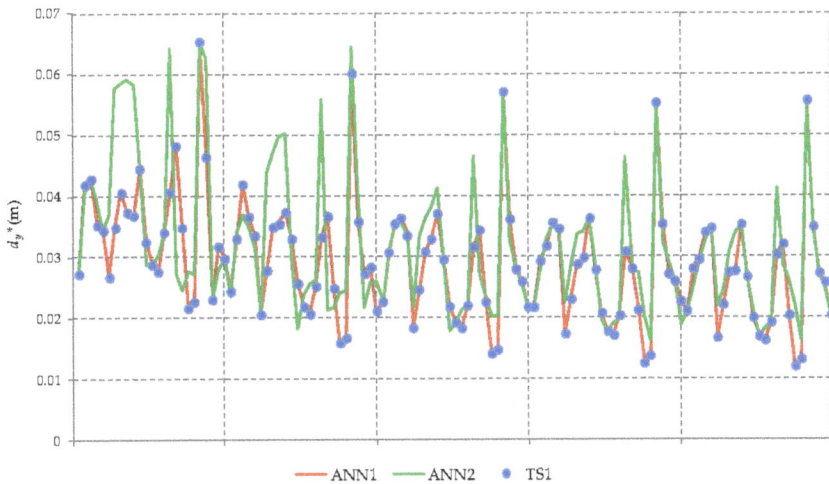

Figure 9. Results comparison for the $d_y{}^*$ values.

It is evident that the results presented in Figure 6 are highly variable, due to the highly nonlinear structural behaviour of the studied buildings, so it seems that the use of a mean capacity curve is not the best approach to assess the seismic vulnerability of this typology.

This is probably why it is so difficult to predict the seismic response of a building when using much more simplified approaches, which are normally used in large-scale studies. The use of ANN may be a valid alternative, if the training sets are representative enough of the problem domain.

Observing the results, it is possible to notice that the ANN1 can reproduce outputs in good agreement with the structural analysis results. However, that is not the case of the ANN2, which are only able to match the results for the 27 training set points (TS2).

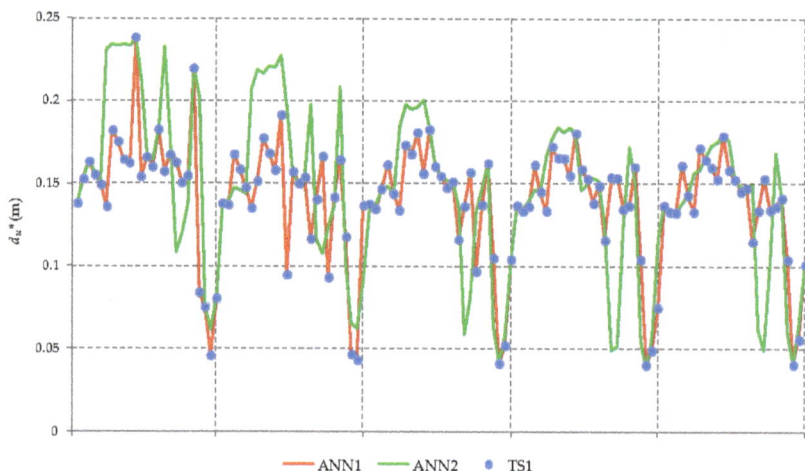

Figure 10. Results comparison for the $d_u{}^*$ values.

To better compare the performance of each neural network, a one-way classical ANOVA F-test [31] was used and the results are presented in Tables 1–3. The lower the F-test results are the higher is the confidence in the ANN results.

Table 1. Classical ANOVA F-test results obtained for $F_y{}^*$ values.

ANN	TS1	TS2
ANN1	0.00004	0.00128
ANN2	1.58284	0.00014

Table 2. Classical ANOVA F-test results obtained for $d_y{}^*$ values.

ANN	TS1	TS2
ANN1	0.0056	0.00113
ANN2	3.26064	0.00007

Table 3. Classical ANOVA F-test results obtained for $d_u{}^*$ values.

ANN	TS1	TS2
ANN1	0.00044	0.00027
ANN2	1.21626	0.00012

The obtained F-test results indicate that ANN2 presents better results than ANN1 when only considering the TS2 capacity curves (which were the 27 capacity curves used to train the ANN2) but ANN1 still presents very good results, because F-test values are almost zero. If only the results obtained with ANN1 and ANN2 were compared with each other for the TS2 case, it would lead to the false conclusion that ANN2 has a better performance than ANN1. However, when considering all the 125 capacity curves of the TS1, it is evident that ANN2 is unable to reproduce the entire domain of the problem, because F-test results for the ANN2 are much higher than zero, being even higher than one. Once again, the ANN1 presents very good results, because F-test values are still almost zero.

The maximum percentage of error of each ANN was also determined for the same input data, which corresponds to each one of the 125 analysis cases of the TS1 and it is presented in Table 4.

Table 4. Maximum percentage of error obtained for the TS1.

ANN	$F_y{}^*$	$d_y{}^*$	$d_u{}^*$
ANN1	10.53%	1.87%	1.67%
ANN2	214.94%	83.95%	141.10%

The highest maximum error value was obtained for the $F_y{}^*$, probably because this variable presents a higher range of values. Again, it is possible to conclude that ANN1 exhibits the lowest errors and they seem to be acceptable to use in large-scale studies, namely having in mind the error that should be expected for this type of studies. On the other hand, ANN2 maximum errors seem to be totally unacceptable, because they are much higher then 100%.

It is important to highlight that the only way to significantly reduce the ANN1 and ANN2 errors seems to be increasing the number of training vectors, to better cover the whole domain of the studied problem. Therefore, the ANN1 presents better results than the ANN2 when all the ANN solutions obtained for the 125 cases (TS1) are compared against the nonlinear analysis results.

Finally, the results obtained for the control cases 1 to 3 were compared to the ones obtained with the previously trained neural networks (Tables 5–7).

Table 5. Results obtained for the control case n. 1.

Case	$F_y{}^*$ (kN)	$d_y{}^*$ (m)	$d_u{}^*$ (m)
CC1	120.25	0.02610	0.14662
ANN1	125.74	0.02789	0.09446
ANN2	154.03	0.02865	0.09375

Table 6. Results obtained for the control case n. 2.

Case	$F_y{}^*$ (kN)	$d_y{}^*$ (m)	$d_u{}^*$ (m)
CC2	89.33	0.01977	0.15324
ANN1	93.98	0.02199	0.15977
ANN2	82.62	0.01802	0.14153

Table 7. Results obtained for the control case n. 3.

Case	$F_y{}^*$ (kN)	$d_y{}^*$ (m)	$d_u{}^*$ (m)
CC3	61.74	0.01761	0.12511
ANN1	63.39	0.01989	0.15139
ANN2	53.06	0.02423	0.18966

The ANN1 presents higher errors when considering the control cases (which are not belonging to the training vectors), in comparison to TS1 results. However, the results still seem to be acceptable in terms of $F_y{}^*$, (maximum error of 5.2%) and $d_y{}^*$ (maximum error of 12.9%), namely in the context of the errors that are usually associated with large-scale studies using more simplified empirical methods. The worst result was obtained for the $d_u{}^*$ (maximum error of 35.6%). These errors would probably be reduced if a higher number of training vectors was used.

5. Conclusions

The results obtained in the present study show that it is feasible to use Artificial Neural Networks (ANNs) to compute simplified capacity curves for seismic assessment purposes. However, the results precision is very dependent on the amount of data used to train the ANN. Moreover, it is important to assure that the entire problem domain is very well covered by the training vectors.

Funding: This research received no external funding.

Conflicts of Interest: The author declares no conflict of interest.

References

1. Calvi, G.M.; Pinho, R.; Magenes, G.; Bommer, J.J.; Restrepo-Vélez, L.F.; Crowley, H. Development of seismic vulnerability assessment methodologies over the past 30 years. *ISET J. Earthq. Technol.* **2006**, *43*, 75–104.
2. FEMA. *HAZUS-MH MR4—Multi-Hazard Loss Estimation Methodology*; Earthquake Model, Technical Manual; Federal Emergency Management Agency: Washington, DC, USA, 2003.
3. Corbane, C.; Hancilar, U.; Ehrlich, D.; De Groeve, T. Pan-European seismic risk assessment: A proof of concept using the Earthquake Loss Estimation Routine (ELER). *Bull. Earthq. Eng.* **2017**, *15*, 1057–1083. [CrossRef]
4. Shustanov, A.; Yakimov, P. CNN Design for Real-Time Traffic Sign Recognition. *Procedia Eng.* **2017**, *201*, 718–725. [CrossRef]
5. Rajasekaran, S.; Febin, M.F.; Ramasamy, J.V. Artificial fuzzy neural networks in civil engineering. *Comput. Struct.* **1996**, *61*, 291–302. [CrossRef]
6. Estêvão, J.M.C. Aplicações da inteligência artificial na engenharia sísmica [Applications of Artificial Intelligence in earthquake engineering]. *Tecnovisão* **2000**, *10*, 17–21. (In Portuguese)
7. Abd-Elhamed, A.; Shaban, Y.; Mahmoud, S. Predicting Dynamic Response of Structures under Earthquake Loads Using Logical Analysis of Data. *Buildings* **2018**, *8*, 61. [CrossRef]
8. Gholizadeh, S.; Salajegheh, J.; Salajegheh, E. An intelligent neural system for predicting structural response subject to earthquakes. *Adv. Eng. Softw.* **2009**, *40*, 630–639. [CrossRef]
9. Lagaros, N.D.; Papadrakakis, M. Neural network based prediction schemes of the non-linear seismic response of 3D buildings. *Adv. Eng. Softw.* **2012**, *44*, 92–115. [CrossRef]
10. Tesfamaraim, S.; Saatcioglu, M. Seismic Risk Assessment of RC Buildings Using Fuzzy Synthetic Evaluation. *J. Earthq. Eng.* **2008**, *12*, 1157–1184. [CrossRef]
11. Morfidis, K.; Kostinakis, K. Seismic parameters' combinations for the optimum prediction of the damage state of R/C buildings using neural networks. *Adv. Eng. Softw.* **2017**, *106*, 1–16. [CrossRef]
12. Ferreira, T.M.; Estêvão, J.M.C.; Maio, R.; Vicente, R. The use of artificial neural networks to assess seismic damage in traditional masonry buildings. In Proceedings of the 16th European Conference on Earthquake Engineering, Thessaloniki, Greece, 18–21 June 2018; pp. 1–10.
13. Morfidis, K.; Kostinakis, K. Approaches to the rapid seismic damage prediction of r/c buildings using artificial neural networks. *Eng. Struct.* **2018**, *165*, 120–141. [CrossRef]
14. Calabrese, A.; Lai, C.G. Fragility functions of blockwork wharves using artificial neural networks. *Soil Dyn. Earthq. Eng.* **2013**, *52*, 88–102. [CrossRef]
15. Ferrario, E.; Pedroni, N.; Zio, E.; Lopez-Caballero, F. Bootstrapped Artificial Neural Networks for the seismic analysis of structural systems. *Struct. Saf.* **2017**, *67*, 70–84. [CrossRef]
16. Xia, Z.; Quek, S.T.; Li, A.; Li, J.; Duan, M. Hybrid approach to seismic reliability assessment of engineering structures. *Eng. Struct.* **2017**, *153*, 665–673. [CrossRef]
17. Vazirizade, S.M.; Nozhati, S.; Zadeh, M.A. Seismic reliability assessment of structures using artificial neural network. *J. Build. Eng.* **2017**, *11*, 230–235. [CrossRef]
18. Alvanitopoulos, P.F.; Andreadis, I.; Elenas, A. Neuro-fuzzy techniques for the classification of earthquake damages in buildings. *Measurement* **2010**, *43*, 797–809. [CrossRef]
19. Fajfar, P.; GašPerŠÍČ, P. The N2 method for the seismic damage analysis of RC buildings. *Earthq. Eng. Struct. Dyn.* **1996**, *25*, 31–46. [CrossRef]
20. CEN. *Eurocode 8, Design of Structures for Earthquake Resistance-Part 1: General Rules, Seismic Actions and Rules for Buildings*; EN 1998-1:2004; Comité Européen de Normalisation: Madrid, Spain, 2004; p. 229.
21. Waszczyszyn, Z.; Ziemiański, L. Neural Networks in the Identification Analysis of Structural Mechanics Problems. In *Parameter Identification of Materials and Structures*; Mróz, Z., Stavroulakis, G., Eds.; Springer: Vienna, Austria, 2005; Volume 469, pp. 265–340.
22. Arslan, M.H. An evaluation of effective design parameters on earthquake performance of RC buildings using neural networks. *Eng. Struct.* **2010**, *32*, 1888–1898. [CrossRef]

23. Estêvão, J.M.C. Modelo Computacional de Avaliação do Risco Sísmico de Edifícios [Computer Model for Buildings Seismic Risk Assessment]. Ph.D Thesis, Instituto Superior Técnico, UTL, Lisbon, Portugal, 1998. (In Portuguese)

24. Estêvão, J.M.C.; Carvalho, A. The role of source and site effects on structural failures due to Azores earthquakes. *Eng. Fail. Anal.* **2015**, *56*, 429–440. [CrossRef]

25. Castaldo, P.; Palazzo, B.; Alfano, G.; Palumbo, M.F. Seismic reliability-based ductility demand for hardening and softening structures isolated by friction pendulum bearings. *Struct. Control Health Monit.* **2018**, *25*, e2256. [CrossRef]

26. Maio, R.; Estêvão, J.M.C.; Ferreira, T.M.; Vicente, R. The seismic performance of stone masonry buildings in Faial island and the relevance of implementing effective seismic strengthening policies. *Eng. Struct.* **2017**, *141*, 41–58. [CrossRef]

27. RBA. *Regulamento do Betão Armado*; Decreto No. 25:948 de 16 de Outubro de 1935; Imprensa Nacional: Lisbon, Portugal, 1935; p. 97. (In Portuguese)

28. Seismosoft SeismoStruct 2016 Release-1—A Computer Program for Static and Dynamic Nonlinear Analysis of Framed Structures. Available online: http://www.seismosoft.com (accessed on 29 July 2017).

29. Boulifa, R.; Samai, M.L.; Benhassine, M.T. A new technique for studying the behaviour of concrete in shear. *J. King Saud Univ. Eng. Sci.* **2013**, *25*, 149–159. [CrossRef]

30. Ojha, V.K.; Abraham, A.; Snášel, V. Metaheuristic design of feedforward neural networks: A review of two decades of research. *Eng. Appl. Artif. Intell.* **2017**, *60*, 97–116. [CrossRef]

31. Cuevas, A.; Febrero, M.; Fraiman, R. An anova test for functional data. *Comput. Stat. Data Anal.* **2004**, *47*, 111–122. [CrossRef]

buildings

MDPI

Article

Retrofitting of Imperfect Halved Dovetail Carpentry Joints for Increased Seismic Resistance

Miloš Drdácký and Shota Urushadze *

Institute of Theoretical and Applied Mechanics of the Czech Academy of Sciences, 190 00 Prague, Czech Republic; drdacky@itam.cas.cz
* Correspondence: urushadze@itam.cas.cz; Tel.: +420-225-443-266

Received: 27 December 2018; Accepted: 13 February 2019; Published: 18 February 2019

Abstract: This paper presents possibilities for anti-seismic improvement of traditional timber carpentry joints. It is known that the structural response of historical roof frameworks is highly dependent on the behavior of their joints, particularly, their capacity for rotation and energy dissipation. Any strengthening, or retrofitting, approach must take into account conservation requirements, usually expressed as conditions involving minimal intervention. Several retrofitting methods were tested on replicas of historical halved joints within various national and international research projects. The joints were produced with traditional hand tools, and made using aged material taken from a demolished building. The paper presents two approaches, each utilizing different retrofitting technologies that avoid completely dismantling the joint and consequently conserve frame integrity. The energy dissipation capacity is increased by inserting mild steel nails around a wooden pin, and connecting the two parts of the halved joint. In the second case, two thin plates made of a material with a high friction coefficient are inserted into the joint and fastened to the wooden elements. This is done by removing the wooden connecting pin and slightly opening a slot for the plates between the halved parts. In addition, the paper presents an application for disc brake plates, as well as thin plates made of oak.

Keywords: carpentry halved joint; energy dissipation; seismic retrofitting

1. Introduction

The behavior of carpentry joints in historical timber structures plays an important role in their overall structural response to applied loads, especially during seismic events. This has been the subject of several research projects, mostly in countries with high seismic activity, such as in the Mediterranean region. In these countries, roof framework joints employ a typical birdsmouth connection for joined timber elements, as shown in Figure 1. For example, Parisi and Piazza [1] tested the rotational capability of such carpentry joints retrofitted with various metal connectors. They have also modeled friction joints in traditional timber structures [2]. Parisi and Cordié [3] further studied the elastic rotational stiffness and post-elastic behavior of double-step timber joints, and the reversed birdsmouth configuration.

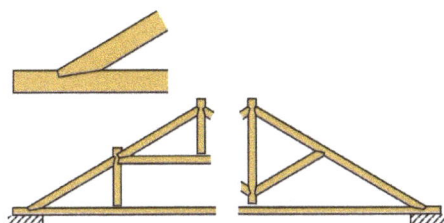

Figure 1. Typical birdsmouth joint and Mediterranean roof structure schemes.

Branco et al. [4] performed a series of tests on joints subjected to static and repeated loading to assess the impact of various strengthening methods. Palma et al. [5] published an extensive review of references, as well as the results of their own experimental research in rotational behavior of rafter and tie beam connections, including a study on the efficiency of some typical strengthening techniques. Similarly, Poletti et al. [6] studied traditional timber joints under cyclic loading, and possibilities of their repair or strengthening. All these works, for the most part, present techniques that are barely applicable to the structural restoration of historic timber structures.

Low slope roofs are typical in the Mediterranean region, in contrast to Central and Northern Europe, where steeper roofs are common and the roof framework structures use different details, especially halved dovetail joints with a higher rotational stiffness. Real joints in historical structures have many imperfections, for reasons ranging from low-quality carpentry work to material defects and degradation. These imperfections substantially influence the stiffness of roof trusses [7], decrease the safety of historical structures, and worsen the response of the roof frames in the event of seismic loading. A study of ways to improve the energy dissipation of carpentry joints was, therefore, carried out within the EC 7th Framework Programme NIKER project, which aimed to establish new integrated knowledge-based approaches for the protection of cultural heritage from earthquake-induced risks.

The performance of roof framework systems under seismic or generally dynamic loading has scarcely been studied even though roofs are sensitive to horizontal loads, especially those perpendicular to the plane of the trussed frame [8], despite the importance of such research having been underlined [9]. The stable roof framework as a three-dimensional box, and its potential role in a seismic event, has been analyzed by Giurani and Marini [10] but, again, only for low slope roofs. Steep roofs have not usually been considered for seismic loadings. However, they are typically constructed in a much stiffer way than low slope roofs, and do influence the building's performance. Their role in this respect has not been appropriately studied, and this is also not an aim of this paper.

With regard to historical structures, conservation requirements constrain retrofitting approaches to minimize interventions. Intervention should not significantly alter the appearance and behavior of a structure, and solutions that do not involve the complete disassembly of existing structures are preferred. This is important in the case of structural restoration works.

During an experimental campaign, a number of retrofitting processes were adopted and tested on replicas of historical halved joints that were made from old wood taken from a demolished building and produced using traditional carpenters' tools and techniques. Presented, here, are two approaches that each utilized a different retrofitting technique.

2. Experiments

2.1. Test Specimens

The experiments were carried out on replicas of traditional halved dovetail joints made from authentic timber (approximately 300 years old) taken from a demolished ancient building. The wooden elements had not been attacked by wood-destroying fungi or by wood-boring insects. Only drying fissures and cracks were present.

2.1.1. Material Characteristics

After the experiments on the joints were complete, the material characteristics of the spruce joint were determined by testing the standard coupons taken from the joint assemblies. The material properties evaluated were wood density and strength properties. Material information, mainly the coefficients of friction of the plates used for increasing energy dissipation, was taken from the literature [11]. The mechanical quantities of the brake plates were taken from the manufacturer [12]. The mechanical properties of the materials are summarized in Table 1.

Table 1. Material characteristics.

Mechanical Property	Spruce	Oak	Brake Plate
Density—ρ (kg/m^3)	340–450	700–750	1950
Compression parallel to grain—R (MPa)	15.63	–	–
Compression perpendicular to grain—R (MPa)	2.09	–	–
Modulus of elasticity—E (MPa)	–	12,000	–
Strength in compression—f (MPa)	–	–	180
Coefficient of friction—μ	–	0.41 *	0.4

* Friction coefficient of the oak parallel to the grain 0.48, perpendicular to the grain 0.34.

2.1.2. Geometrical Characteristics

The geometry of the joint assemblies varied, as did the extent of the wood's deterioration. An angle of 45° between the joined members was adopted. This is close to the angles most widely used in Central European roofing frames. The geometry of the specimens, the locations of the acting forces, and the potentiometer are depicted in Figure 2. The specimens were produced with intentional imperfections, as far as the tightness of the coupled elements was concerned, to model some degree of joint degradation. For this research, only two types of imperfection were considered—perfect joints with a tight connection between joined short struts in the overlap, and imperfect joints exhibiting a slight loss of contact in the abutment of the joined elements. The imperfections were produced during high-quality carpentry work and, therefore, their range was limited to local defects of 3 mm maximum. The geometric imperfections exhibit rather high slippage during the change in direction of the loaded arm rotation, as shown in Figure 8 and Figure 9, compared to a perfectly made joint, e.g., as shown in Figure 13. The increase in overall deformation increment of a roof truss due to joint slippage decreases with an increase of slippage value. This dependence is not linear, and the greatest difference in the overall deformation has been analyzed only around the change from perfect to imperfect joint geometry [7].

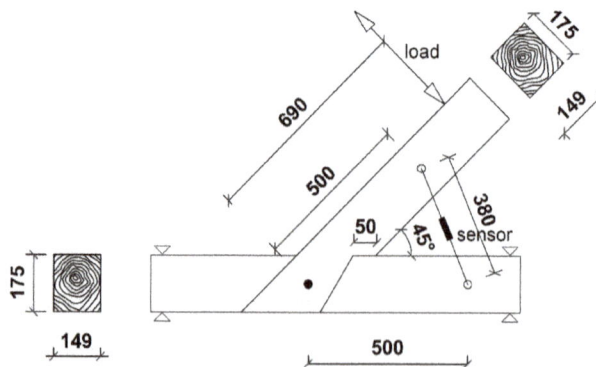

Figure 2. Geometrical scheme of the joint, the potentiometer location and location of the acting force, the direction of the joint's rotation, and the corresponding moment.

2.2. Joint Retrofitting

Several retrofitting techniques for increased energy dissipation were suggested and tested, as described above. However, for historic roof frameworks, only those which were acceptable from an aesthetic point of view were relevant for further development, especially in cases where the historic structures were exposed and accessible to visitors. Furthermore, the requirements of minimum intervention, re-treatability, and easy implementation were decisive for the selection of the tested techniques. Only the two most-suitable approaches for practical applications will be presented, as follows.

2.2.1. Metal Nails around a Wooden Pin

In the first case, energy dissipation capacity was increased by inserting mild steel rods (nails) around the wooden pin that connects the two parts of the halved joint, as shown in Figure 3a,b. No dismantling was necessary for this retrofitting intervention. The 6 mm diameter (D) nails were applied into pre-bored holes, which allowed for minimum spacing. When using six nails, the standard recommendations (10 × D along grains, 3 × D across grains) were observed. When using eight nails, due to the grain inclination, the distance along the grain was only 6 × D.

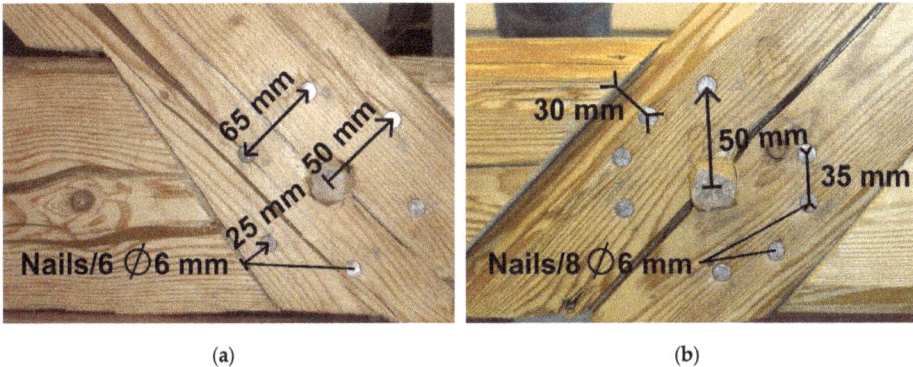

Figure 3. (**a**) Retrofitting using six mild steel nails; (**b**) retrofitting using eight steel nails.

2.2.2. Friction Joints

In the second retrofitting approach, the connecting wooden pin was removed from the joint, the halved parts were slightly separated, and two thin plates were then inserted into the opened slot and fastened to the wooden elements, as shown in Figures 4 and 5a,b. The plates were made of material with a high friction coefficient. Car disc brake plates and thin oak plates were used. The joint was then fixed and tightened with a steel bolt, which was pre-stressed to a certain degree. The screw bolt enables the joint's pre-stress level to change, which influences not only the stiffness of the joint, but also the frictional force between the plates. Several bolt pre-stressing moment values were chosen and tested (from 115 up to 240 Nm), which generated a stress level on the friction surfaces of around 0.43–0.90 MPa.

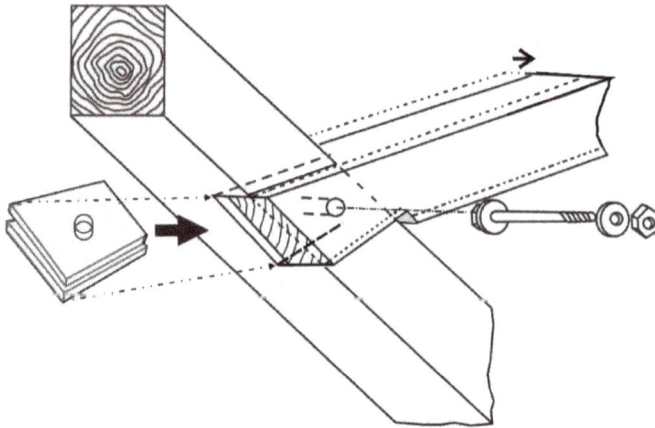

Figure 4. Schematic sketch showing insertion of friction plates into a halved dovetail joint, without complete dismantling, which is the main advantage of this approach.

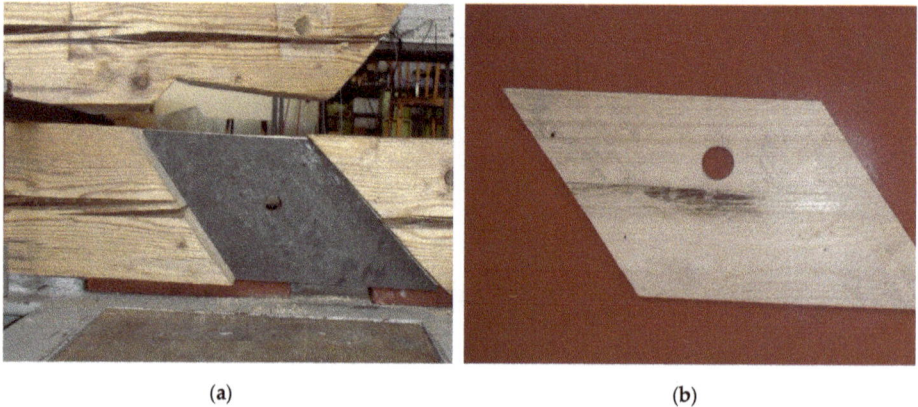

(a)   (b)

Figure 5. (a) Joint retrofitting with brake friction plate and replacement of the oak pin by a pre-stressed steel screw bolt; (b) an oak plate for joint retrofitting with friction plates.

2.3. Test Arrangement

On the basis of positive experience gained from previous tests performed at the Institute of Theoretical and Applied Mechanics of the Czech Academy of Sciences (ITAM) [7,13], a similar test set-up was constructed, as shown in Figure 6. This set-up enabled cyclic loading and pseudo-dynamic behavior of the halved dovetail joints to be simulated separately from the roof frame. The joint samples for testing were placed into a special testing rig that enabled pseudo-dynamic cyclic loading. It also ensured static stability of the samples and their responses in only the direction of loading. The cyclic load was applied using a servohydraulic MTS Systems Corporation actuator (cylinder) (MTS headquarters, 14000 Technology Drive, Eden Prairie, MN, USA, 55344) with a capacity of 25 kN, attached to a steel frame. The rotational responses of the joints were measured indirectly by means of a MEGATRON Elektronik GmbH & Co. KG SPR 18-S-100 (5 kΩ) potentiometer (MEGATRON Elektronik GmbH & Co. KG • Hermann-Oberth-Strasse 7 • 85640 Putzbrunn/Munich, Germany).

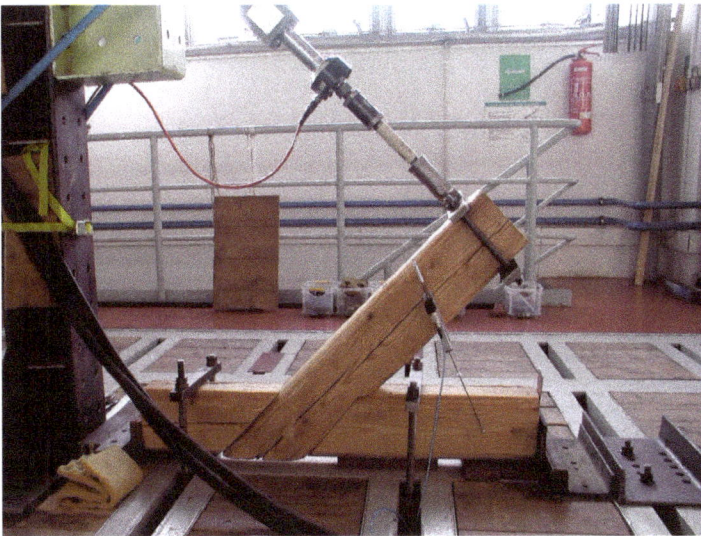

Figure 6. The test set-up.

The specimens were cyclically loaded, and the load deformation curves were registered. The load was applied to the joint using the actuator attached to the oblique beam. The intensity of the actuator's force was controlled by its prescribed displacement. The amplitude of the controlled displacement increased for each cycle, with a constant step equal to 4 mm. The frequency of each cycle was 0.1 Hz. During the tests, the forces needed to achieve the desired displacement of the cylinder and the change in the length of the potentiometer were recorded, as shown in Figure 7.

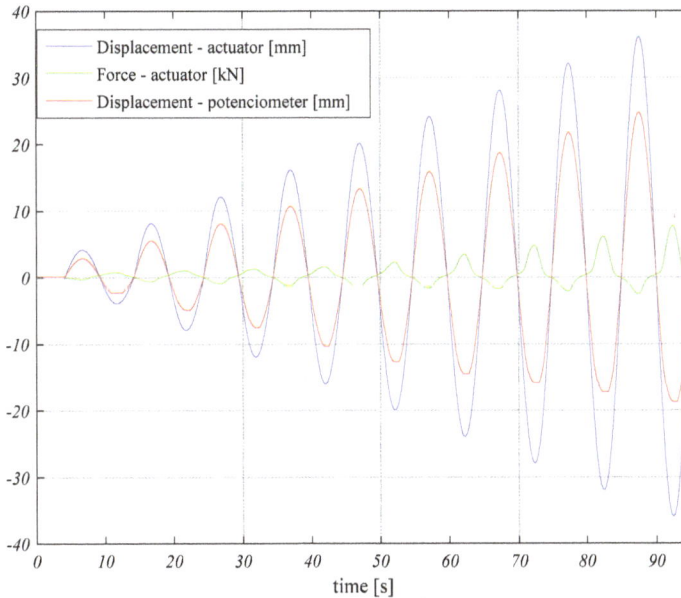

Figure 7. Time histories of the measured quantities during cyclic tests.

3. Results

Damping property is a fundamental factor influencing a structure's seismic resistance. Effective dissipation of energy by a structural element, e.g., a joint, can significantly decrease a structure's vibration level and lessen internal forces. The dissipative properties of the investigated halved dovetail joints can be described by the area of hysteresis loops, as shown in Figures 8a,b,9–11. Here, the hysteresis loops area representation of how the actuator's moment of force about the axis of pin M is dependent on the rotation of joint Δα for one loading cycle.

Both of the retrofitting methods tested (see Discussion) were effective from an energy consumption point of view. The results showing the changes in hysteresis loops after joint retrofitting are presented in Figures 8a,b,9–11. Figure 8b clearly shows an apparent increase in the energy needed to achieve the required rotation (displacement), in contrast to the unreinforced joint, represented in Figure 8a. In this case, the cyclic loading continued after the maximum testing load was reached, following the stability of the response loop. A negligible change in the hysteresis loop was observed.

Similar positive effects were attained in tests with the inserted friction plates. A typical example of the behavior of a joint with two brake plates inserted in the gap between overlapping parts of the joined elements is shown in Figure 10. As shown in Figure 11, an almost identical dissipation energy was achieved with the two oak plates (Figure 11), even though the pre-stressing forces applied to the joining bolts were not identical.

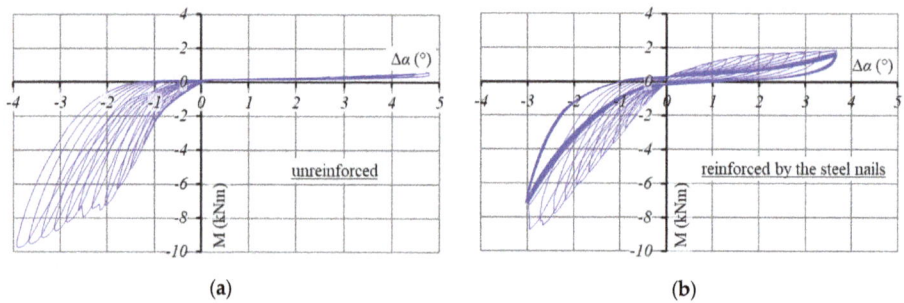

(a)   (b)

Figure 8. (a) Hysteresis loops of a joint before retrofitting with steel nails; **(b)** Hysteresis loops of a joint after retrofitting with steel nails.

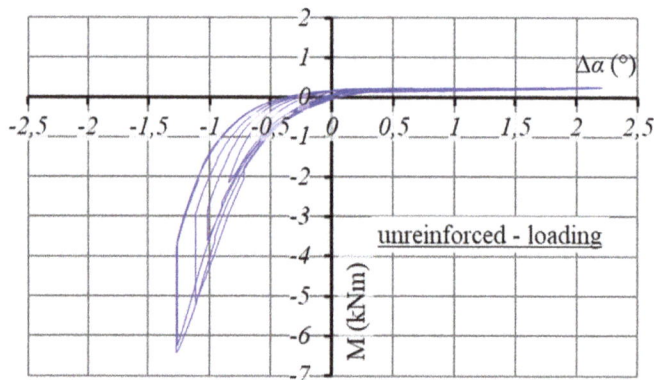

Figure 9. Hysteresis loops of a typical imperfect halved dovetail joint before retrofitting.

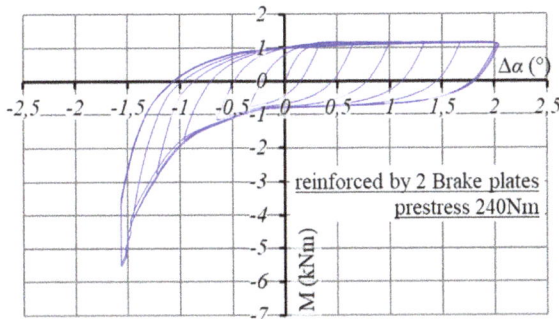

Figure 10. Hysteresis loops of a joint retrofitted with brake plates.

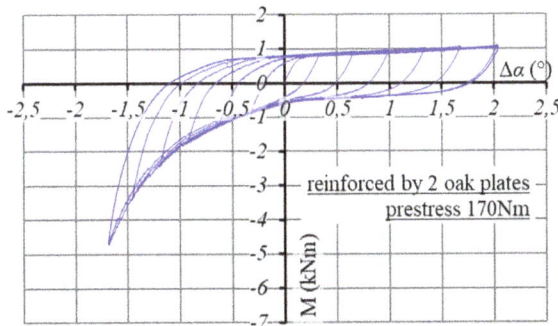

Figure 11. Hysteresis loops of a joint retrofitted with oak plates.

In the friction joints, energy dissipation depended on the forces pressing the two surfaces together. The resulting increase in energy dissipation, during cycling, of the tested joint retrofitting variations is shown in Figure 12. Comparing the results of the effect of two oak plates in relation to the brake plates—at an identical pre-stressing force—showed an obviously higher efficiency for the brake plates.

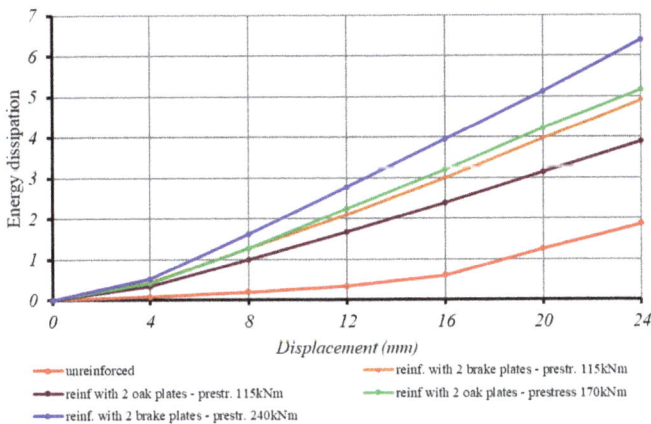

Figure 12. Energy dissipation curves for joints, both unreinforced and reinforced with various types of structural modifications.

4. Discussion

Rotational stiffness and slipping of joints were studied in detail by Drdácký et al. [7] for both a typical baroque and gothic roof framework on which halved dovetail joints were widely used. The results showed that tight, well-made joints with a reduced chance of slipping decreased the overall roof frame deformation under horizontal wind or earthquake loads, while also providing reasonably high rotational stiffness and capacity. Highly skilled carpenters were capable of producing flawless joints which were free of gaps between the individual elements.

Perfectly made joints may exhibit a sufficiently high rotational capacity, as shown in Figure 13, and as also shown by Wald et al. [13], and can be modeled in a rather simple way using a component modeling approach, e.g., after Vergne [14], which represents a good tool for the description joint behavior under repeated loading, as shown in Figure 14. In such a model, the joint was divided into components, which were represented by a force–deformation diagram. It is supposed that the dowel resists the shear force and clearly fixes the position of the connection's center of rotation. The results presented in Figure 14 show that more testing of the materials' characteristics are required, especially the behavior of wood under concentrated compression, to be able to describe the contact in the rafter dovetail indent. The tested composed joints are more complex and, therefore, an experimental investigation was adopted without any attempt to create computational models.

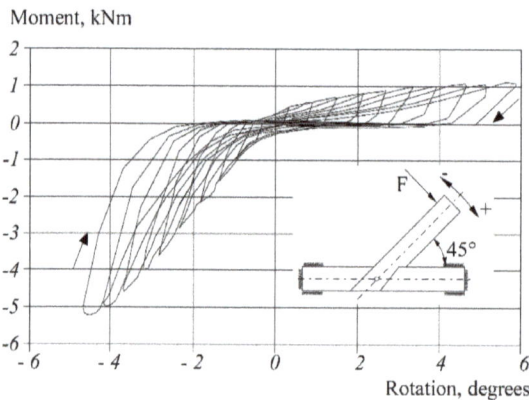

Figure 13. Example of an experimentally attained moment rotation diagram of a perfectly made halved dovetail joint [7].

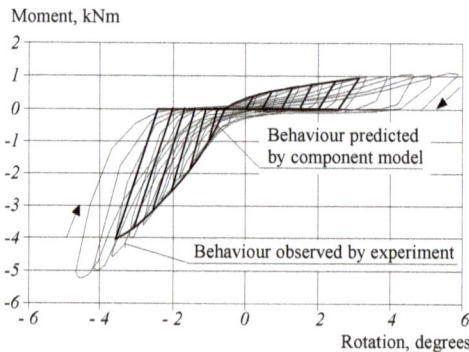

Figure 14. Comparison of the predicted component-based model for cyclic loading in the moment rotation diagram in Figure 13 [7].

Inserting nails around a wooden pin is a modification of the modern design of seismically resistant frame corners, Ceccotti A. [15], Kasal et al. [16]. The positive effect of this design has been published (for an example, Stehn and Börjes [17]). Our experiments have also demonstrated the effectiveness of this design, as shown in Figure 15a,b. The nails additionally strengthen and integrate the joint, and may replace a degraded wooden pin.

As the timber structure moves, the connected timber struts rotate mutually in the joint, and tend to deform the inserted steel rods plastically, which absorbs the energy, as shown in Figure 15b,c.

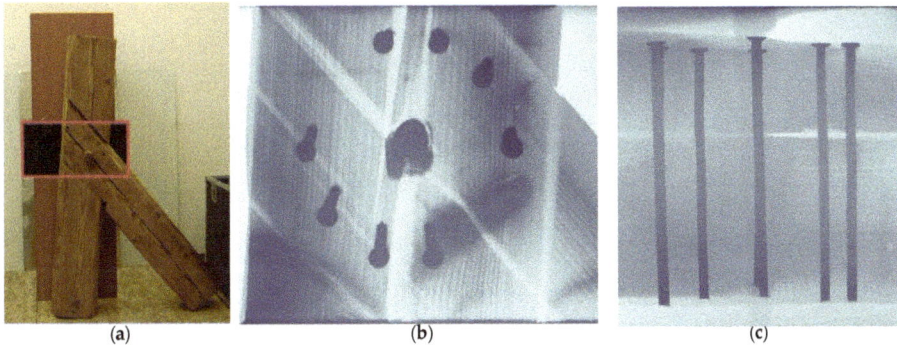

(a)   (b)   (c)

Figure 15. X-ray presentation of the deformed mild steel nails after the cycling tests of a joint: the (**a**) X-ray recording, (**b**) perpendicular view of the joint plane, and (**c**) cross-sectional view of the joint plane.

A rather complex design, supported by appropriate numerical modeling, is required for an optimal interaction between nails and wood. The effect can be improved by inserting and fixing fiber-reinforced plastic (FRP) sheets on the contact surfaces of the halved joint counterparts. This requires partial dismantling of the joint, but prevents the origination and propagation of fissures or cracks.

Similarly, the approach based on improvement of friction between the contact surfaces of the halved joint requires moderate intervention, i.e., temporary removal of the wooden joining pin, and insertion and fixation of thin plates in the slot between the two joint components. The results are presented for an unreinforced joint, as shown in Figure 8; for brake plates, as shown in Figure 10; and for oak plates, as shown in Figure 11.

5. Conclusions

This experimental investigation focused on determining the effectiveness of several joint modifications with respect to their dissipative properties. The best results, and the highest level of effectiveness, were obtained for joints combined with plates having a high friction coefficient and a steel bolt. Tightening the nut on the bolt that is replacing the pin influences the pre-stress of the joint, i.e., the friction force between the plates. The frictional force increases in direct proportion to an increase in the degree of pre-stress. However, compressive deformation of the wood limits the maximum possible value of friction force. Fully fastening the brake plates to the wood using screws was the most effective method. Plates made of oak can be used as an alternative to brake plates, however, their damping efficiency is not as high. Reinforcing the joints with nails gave good results, as did the use of a combination of nails and inserted plates.

Author Contributions: M.D. and S.U. have done research and analyses, interpreted the results, as well as writes all sections of the paper. Original Draft Preparation, Writing-Review & Editing, M.D.

Funding: This study was supported by the EC FP7 Collaborative NIKER project (Grant agreement No. 244123) [18,19], the grant project DG18P02OVV040 "Monuments in motion", NAKI program II, provided by the Ministry of Culture of the Czech Republic and the ITAM institutional fund RVO 68378297.

Acknowledgments: The writers would like to thank master carpenter Stejskal for careful preparation of the test specimens, to Jaroslav Valach for X-ray photography and Marek Eisler for language and style correction.

Conflicts of Interest: The authors declare no conflict of interest.

References

1. Parisi, M.A.; Piazza, M. Seismic behavior and retrofitting of joints in traditional timber roof structures. *Soil Dyn. Earthq. Eng.* **2002**, *22*, 1183–1191. [CrossRef]
2. Parisi, M.A.; Piazza, M. Dynamic modeling of friction joints in traditional timber structures. In Proceedings of the Eurodyn'99 Conference, Prague, Czech Republic, 7–10 June 1999.
3. Parisi, M.A.; Cordié, C. Mechanical behavior of double-step timber joints. *Constr. Build. Mater.* **2010**, *24*, 1364–1371. [CrossRef]
4. Branco, J.M.; Piazza, M.; Cruz, P.J.S. Experimental evaluation of different strengthening techniques of traditional timber connections. *Eng. Struct.* **2011**, *33*, 2259–2270. [CrossRef]
5. Palma, P.; Garcia, H.; Ferreirac, J.; Appletond, J.; Cruz, H. Behaviour and repair of carpentry connections—rotational behaviour of the rafter and tie beam connection in timber roof Structures. *J. Cult. Heritage* **2012**, *13S*, S64–S73. [CrossRef]
6. Poletti, E.; Vasconcelos, G.; Branco, J.M. Performance evaluation of traditional timber joints under cyclic loading and their influence on the seismic response of timber frame structures. *Constr. Build. Mater.* **2016**, *127*, 321–334. [CrossRef]
7. Drdácký, M.; Wald, F.; Sokol, Z. Sensitivity of Historic Timber Structures to Joint Response. In Proceedings of the 40th Anniversary Congress of IASS Madrid, Madrid, Spain, 20–24 September 1999; Astudillo, R., Madrid, A.J., Eds.; 1999; Volume 2, pp. G1–G10.
8. Parisi, M.A.; Chesi, C.; Tardini, Ch. Inferring Seismic Behavior from Morphology in Timber Roofs. *Int. J. Archit. Heritage* **2012**, *6*, 100–116. [CrossRef]
9. Parisi, M.A.; Piazza, M. Seismic strengthening and seismic improvement of timber structures. *Constr. Build. Mater.* **2015**, *97*, 55–66. [CrossRef]
10. Giuriani, E.; Marini, A. Wooden roof box structure for the anti-seismic strengthening of historic buildings. *Int. J. Archit. Heritage* **2008**, *2*, 226–246. [CrossRef]
11. Aira, J.R.; Arriaga, G.; Íñiguez-González, G.; Crespo, J. Static and kinetic friction coefficients of Scots pine (Pinus sylvestris L.), parallel and perpendicular to grain direction. *Materiales de Construcción* **2014**, *64*, 1–9. [CrossRef]
12. Available online: http://www.engineershandbook.com/Tables/frictioncoefficients.htm (accessed on 20 December 2018).
13. Wald, F.; Mareš, J.; Sokol, Z.; Drdácký, M. Component method for historical timber joints. In *The Paramount Role of Joints into the Reliable Response of Structures*; NATO Science Series; Baniotopoulos, C.C., Wald, F., Eds.; Kluwer Academic Publishers: Dordrecht, The Netherlands, 2000; pp. 417–424.
14. Vergne, A. Testing and modelling of load carrying behaviour of timber joints. In Proceedings of the International Conference Control of Semi-Rigid Behaviour of Civil Engineering Structural Connections, Liege, Belgium, 17–19 September 1998; p. 4.
15. Ceccotti, A. Structural Timber: Characteristics and Testing. *Struct. Eng. Int.* **1993**, *3*, 95–98. [CrossRef]
16. Kasal, B.; Pospíšil, S.; Jirovský, I.; Heiduschke, A.; Drdácký, M.; Haller, P. Seismic performance of laminated timber frames with fiber-reinforced joints. *J. Earthq. Eng. Struct. Dyn.* **2004**, *33*, 633–646. [CrossRef]
17. Stehn, L.; Börjes, K. The influence of nail ductility on the load capacity of a glulam truss structure. *Eng. Struct.* **2004**, *26*, 809–816. [CrossRef]

18. D 6.4 (2010) Experimental Result on Structural Connections. Deliverable 6.4, EC FP7 Collaborative Project NIKER—New Integrated Knowledge Based Approaches to the Protection of Cultural Heritage from Earthquake Induced Risk (Grant Agreement No.: 244123). Available online: www.niker.eu/ (accessed on 20 December 2018).
19. Drdácký, M.; Urushadze, S.; Wünsche, M. Retrofitting of imperfect carpentry joints for increased seismic resistance. In *SAHC 2012 "Structural Analysis of Historic Constructions"*; Jasienko, J., Ed.; Dolnoslaskie Wydawnictwo Edukacyjne: Wroclaw, Poland, 2012; pp. 1485–1492.

buildings

MDPI

Article

Seismic Vulnerability for RC Infilled Frames: Simplified Evaluation for As-Built and Retrofitted Building Typologies

Marco Gaetani d'Aragona *, Maria Polese, Marco Di Ludovico and Andrea Prota

Department of Structures for Engineering and Architecture, University of Naples Federico II, 80125 Naples, Italy; maria.polese@unina.it (M.P.); marco.diludovico@unina.it (M.D.L.); andrea.prota@unina.it (A.P.)
* Correspondence: marco.gaetanidaragona@unina.it

Received: 7 August 2018; Accepted: 26 September 2018; Published: 28 September 2018

Abstract: Several studies investigated the influence of infills on the response of reinforced concrete (RC) frames. However, possible shear brittle failures are generally neglected. The interaction between the infill panels and the surrounding frames can lead to anticipated brittle-type failures that should be considered in code-based assessment of lateral seismic capacity. This paper investigates, by means of simplified pushover analyses, on the effect of infills on the lateral seismic capacity explicitly considering possible brittle failures in unconfined beam-column joints or in columns. Archetype buildings representative of existing gravity load designed (GLD) RC frames of three different height ranges are obtained with a simulated design process and a sensitivity analysis is performed to investigate on the effect of infill consistency on the capacity. Moreover, possible alternative local retrofit interventions devoted to avoiding brittle failures are considered, evaluating their relative efficacy in case of different infill typologies. It is seen that for the considered existing GLD buildings, the attainment of life safety limit state is premature and happens before the damage limitation limit state. The capacity can be increased with application of local retrofit interventions. However, the retrofit efficacy varies depending on the infills consistency if the horizontal action transferred from the infills to the surrounding frame is not absorbed by the retrofit solution.

Keywords: infilled frames; pushover; seismic capacity; frame-infill interaction; retrofit; FRP; brittle failure; joints

1. Introduction

Seismic vulnerability of buildings represents the susceptibility of buildings to be damaged by earthquakes of given intensity. Depending on the scope of the vulnerability analysis, different levels of damage may be considered. Generally, with reference to civil protection purposes, the damage scale is defined by a qualitative description of the damage on structural and nonstructural elements, also considering the damage severity and extent, and refers to an estimation for the entire building. The reference damage scale in Europe is that defined in the European macro-seismic scale EMS98 [1], while indications for United States may be found in [2]. On the other hand, vulnerability studies may be used also with the purpose of evaluating the building seismic capacity corresponding to the attainment of code-based limit states, evidencing possible weaknesses in selected building typologies. In such a case, the damage scale is graduated with reference to selected relevant limit states and often deformation limits at a storey, or the crisis of a single element drives the "failure" of the entire building. The adoption of code-based limits for damage graduation is useful for estimating the possible overpassing of selected limit states relevant for the codes (e.g., damage limitation limit state or life safety limit state), the eventual need to retrofit or the effect of selected mitigation interventions to reduce earthquake damage [3].

This paper performs a vulnerability study on reinforced concrete (RC) infilled frame buildings, representative of existing gravity load designed typologies in Italy.

Infill panels are usually treated as non-structural elements, considering their effect as substantially beneficial with respect to building's seismic behavior [4]. In fact, in the case of uniformly infilled structures, with infill panels having adequate consistency, the buildings are generally less vulnerable with respect to equivalent bare structures [5]. However, the seismic behavior of the frame-infills assemblage is strongly influenced by a number of factors concerning infills strength and distribution: if the infills are not present or weakened due to the presence of large openings at one storey there is a high probability of soft storey effects; this circumstance may occur also for uniformly distributed infills, when they are weak and relatively brittle and/or for high seismic demand with respect to the system resistance [6,7]. Moreover, if the infill contribution in terms of strength and stiffness is not adequately proportioned with respect to the RC frame, the triggering of a number of local effects may affect the formation of a global-type mechanism, leading, instead, to a premature collapse of the columns at a single storey [8]. This is even more probable in case of non-conforming elements with inadequate reinforcement detailing, where local interaction between infills and RC columns may cause structural collapse due to shear failure of the columns or due to a shear friction failure.

Strength and stiffness characteristics of the infill panels, and more generally of the infill-frame assemblage, are strongly dependent on a number of factors such as compressive strength, shear (cracking) resistance, longitudinal and transversal elastic modulus of the masonry panels, the quality of connection between the panels and the surrounding frame, the openings, workmanship ability, etc. [9,10]. For what concerns the mechanical properties, although they may be associated to the characteristics of the natural/artificial masonry elements and of the mortar forming the panel, and on the disposition of the elements (vertical or horizontal rows, overlapping etc.), it is still observed a high variability of these properties, with coefficients of variation that may be major than 40% [11].

Several studies attempted to evaluate the influence of infills in the linear or nonlinear response of RC frames. For example, in [12] a parametric investigation on the effect of the infill wall area on the global elastic drift ratios of RC frames is performed, in [13,14] the variation of elastic drift ratios considering infills distribution and opening percentage is studied, while in [15] a probabilistic analysis considering the effect of uncertainties of the infills elastic modulus and of the equivalent strut's effective width on the building elastic response is presented. However, the presence of infills may alter the RC structural response also in the non-linear stage, e.g., by modification of the type of collapse mechanism [16]. In [17] the seismic reliability on infilled RC frames is studied by means of non-linear dynamic analyses of representative frames, in [18] is evaluated the sensitivity of the seismic response parameters to the uncertain modelling variables of the infills and frame by means of pushover analyses and in [19] a parametric investigation on the results of pushover analyses to evaluate the influence of the mechanical and geometrical properties of masonry infills on the whole structural response is performed. However, these studies do not consider possible brittle failures in unconfined lateral joints or in columns with inadequate transversal reinforcement. The interaction between the infill panels and the surrounding frames can lead to anticipated brittle-type failures that should be considered in code-based assessment of lateral seismic capacity. Additionally, when designing possible local retrofit interventions on nonconforming elements, the effect of the aggravating lateral solicitations transmitted from the infill panel to the RC elements should be properly considered.

This paper investigates on the effect of infills on the lateral seismic capacity considering possible brittle failures in unconfined lateral joints or in columns, which are a common cause of collapse for existing RC buildings not designed for seismic loads or with obsolete seismic provisions. Archetype buildings representative of existing gravity load designed RC frames of three different height ranges are obtained with a simulated design process and a sensitivity analysis is performed to investigate on the effect of infill consistency on the capacity. Moreover, possible alternative local retrofit interventions devoted to avoiding brittle failures are considered, evaluating their relative efficacy in case of weak or stronger infills. Building response is studied performing simplified nonlinear static pushover analyses and evaluating the lateral seismic capacity at code-based damage limitation SLD and life-safety SLV limit states [20,21]. Frame-infill interaction is evaluated by suitable post-processing of pushover results allowing the evaluation of lateral displacement corresponding to the attainment of seismic capacity.

2. Analytical Based Assessment of the Seismic Capacity

The lateral seismic capacity is evaluated by means of nonlinear static pushover, identifying on the pushover curve the points corresponding to the attainment of two limit states, namely damage limitation limit state SLD and life safety limit state SLV. Then, the capacity is expressed in terms of the maximum peak ground acceleration PGA_c that the building is able to withstand for each limit state ($PGA_{c,SLD}$ and $PGA_{c,SLV}$). A capacity spectrum method approach is employed for transformation of the capacity curve relative to the multi-degree of freedom model MDOF representative of the real building to the corresponding equivalent single degree of freedom SDOF system and for evaluation of the capacity with a spectral approach [22].

2.1. Nonlinear Building Modeling

The structural model for the generic archetype building is obtained with a simulated design approach as already described in [23] and further simplified in [24]. In this paper, we consider only gravity load designed (GLD) regular buildings of rectangular shape. In particular, complying with the building codes (e.g., [25,26] and following modifications or integrations) and design practice in force at the time of construction (e.g., [27]), the structural elements are firstly dimensioned; for this application we consider GLD buildings constructed in Italy between 1950 and 1980. The longitudinal reinforcement in columns is designed with reference to minimum longitudinal reinforcement geometric ratio prescribed by code for gravity load designed buildings, while for the beams longitudinal reinforcement is designed considering envelope moments deriving by limit load combination schemes according to construction practice. Note that in the typical structural configuration of gravity load designed buildings, the horizontal slabs are one-way slabs. Hence, the deep beams mainly deputed to absorb gravity loads are just in the direction perpendicular to the one-way slabs. This implies that generally the RC frames formed by columns and deep beams are only in one direction, while in the other direction the frames are formed by columns and beams that are embedded in the thickness of the horizontal slab; only the perimeter frames are always with deep beams [27]. More details for the simulated design process can be found in [23]. The possible presence of infills is also considered. The perimeter frames are infilled, and it is supposed that the area of openings for the infills correspond to 20% of the infill area for each panel.

Thanks to symmetry, each building is analyzed separately in both the longitudinal (X) and the transversal (Y) directions, simply assembling the contribution of the relevant frames as acting in parallel. The nonlinear model includes the effect of the infills, which are modeled as equivalent single strut acting only in compression. Global model is assembled considering that ends of the columns are restrained against rotation (Shear Type model), as already proposed in [28]. This simplifying hypothesis was already introduced in other studies for the evaluation of RC infilled frames [29]; it allows the analyses to reproduce the seismic response of existing buildings with a reasonable degree of approximation and at the same time to reduce the computational effort, which is a crucial aspect when large scale analyses have to be performed. This way, the lateral response of the structure under a given distribution of lateral forces can be determined based on the interstorey shear–displacement relationships at each storey.

Figure 1 shows a schematic representation of the generic frame of the building with identification of the columns and infill panels and the nonlinear model adopted for each of these elements. For RC columns, a tri-linear moment-rotation envelope is built with cracking and yielding as characteristic points (Figure 2). Moment at yielding (M_y) is calculated according to the simplified formulation proposed by Biskinis and Fardis [30], while the rotation at yielding (θ_y) is identified by M_y and the secant stiffness (EI_y) provided by Haselton et al. [31]. The total shear demand acting on the column (black continous line in Figure 2b) is evaluated by transforming the moment-rotation (M-θ) relationship into the corresponding shear-displacement (V-Δ) curve (dashed gray) and summing the effect of infill-induced shear stress (dotted gray). Then, for the non-ductile RC building, the possible shear failure of columns is identified by comparing the obtained column shear demand with the column shear capacity (red continuous line) evaluated as described in Section 2.2. The column shear capacity can be improved by retrofitting the building through the application of external fiber reinforced polymers FRP layers as described in Section 2.4. Figure 2b shows the increased shear capacity when applying one (dashed red) or two (dotted red) FRP plies as a retrofit option. Note that, for the selected column, two FRP plies may avoid column brittle failure.

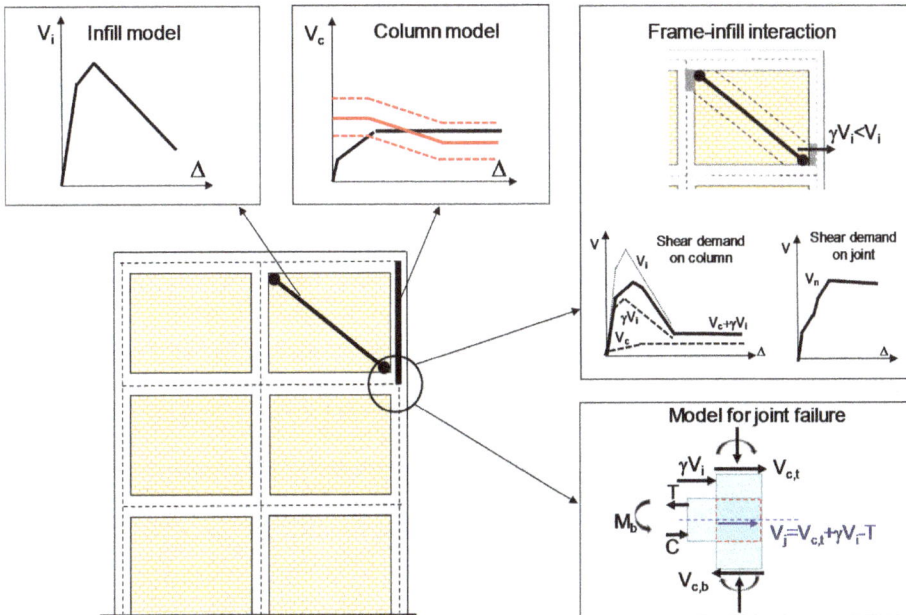

Figure 1. Schematic representation of the generic frame of the building with identification of the columns and infill panels and the nonlinear model adopted for each of these elements. The two panels on the right schematize frame-infill interaction and modeling approach to capture shear demand on columns and joints.

The behavior of infill panels can be simulated adopting either micro or macro-modelling techniques [32]. The latter are frequently adopted due to their simplicity and efficiency into the description of the global response of masonry infilled RC frames. In particular, the single-strut modeling approach is widely adopted to simulate the behavior of global effects of infills on frames. Several models were proposed for the determination of the force-displacement relationship of the equivalent single strut masonry infills (e.g., [32,33]). In this study, the force-displacement envelope of the equivalent diagonal struts is evaluated according to the model proposed by Panagiotakos and Fardis [33], see Figure 3. In this model, the four branches lateral force-displacement relationship is constructed depending on the geometry of the surrounding frame, and on both mechanical and geometrical characteristics of the infill masonry [34].

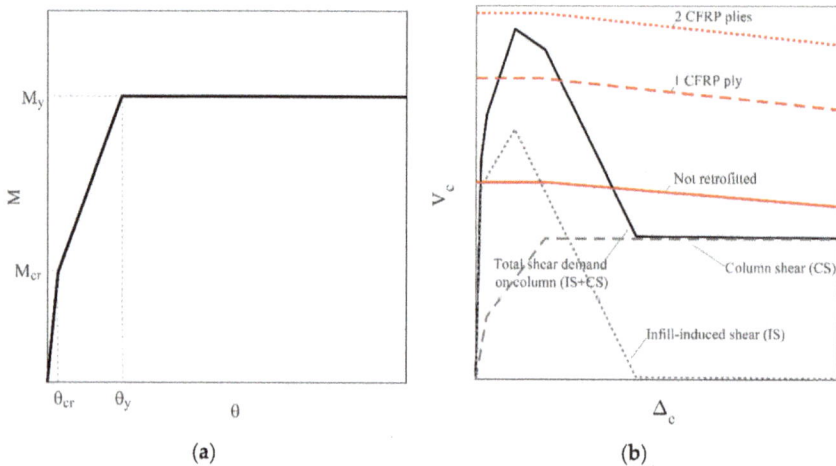

Figure 2. Moment-rotation relationship for RC columns (**a**) and the corresponding V-Δ diagram (**b**). In (**b**) also the additional shear on columns induced by infill panels is shown, along with the adopted shear capacity model for not retrofitted building and for different retrofit options.

The geometry of the infill panel is introduced in terms of gross length and height, L and H, clear length and height, l_w and h_w, clear diagonal length, d_w, and in terms of inclination of the diagonal strut $\theta = \arctan(h_w/l_w)$ with respect to the horizontal plane. The mechanical characteristics of the masonry are expressed in terms of elastic shear modulus, G_w, Young's modulus, E_w, and shear cracking strength, τ_{cr}. Finally, the equivalent strut width, b_w, is determined according to Mainstone's formula [35] depending on quantities reported above and on the moment of inertia, I_c, and Young's modulus, E_c, of columns. When openings are present in the infill panel, e.g., to accommodate windows or balconies, both the stiffness and the strength of the infill panel are reduced. Several studies investigated the influence of shape, size and location of openings on the seismic performance of infilled frames (e.g., [36,37]). In this study the presence of the opening is considered introducing a reduction factor $\lambda_0 = \max(0; 1-1.8\, A_o/A_p)$ [38] that modifies both the stiffness and the strength of the infill panel, where A_o is the area of openings and A_p the area of infill panels.

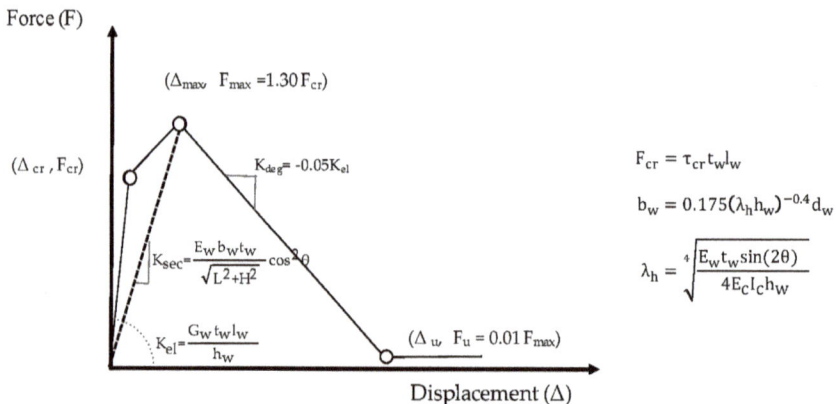

Figure 3. Lateral force–displacement relationship for infill panels.

Once that force-displacement/moment-rotation relationships is defined for each relevant member, these are transformed in the corresponding shear-displacement curves and then, considering the RC frames and infill elements at the same storey as acting in parallel, a multi-linear interstorey shear-displacement relationship is constructed for each storey. Assuming a given lateral force distribution shape, e.g., proportional to first mode shape or with forces proportional to storey masses, the global pushover curve is finally obtained in closed-form through a force-controlled procedure up to the peak response. After the peak, a displacement-controlled procedure is followed, as suggested in [28]. In particular, the storey in which the maximum ratio between interstorey shear demand and strength occurs is the only one to reach its peak interstorey shear resistance and to experience softening post-peak behavior. In this storey, the interstorey displacement continuously increases, while the displacement in remaining storey will decrease following an unloading branch with a stiffness value equal to the elastic one.

2.2. Limit States for Non-Ductile RC Buildings

Two relevant code-based limit states, namely damage limitation (SLD) and life safety limit state (SLV), are detected along the pushover curves to represent lateral seismic capacity of the building.

Concerning SLD, modern codes such as EC8–part 1 [20] and the Italian code [21], prescribe that maximum interstorey drift IDR_{max} should not exceed the value of 5‰ for buildings having brittle non-structural components connected to the structure, e.g., in case of brick masonry. If the contribution of infill panels is explicitly considered in the model the threshold for IDR_{max} should be reduced [39], and a value that refers to ordinary masonry, i.e., 2‰, should be adopted.

Adopting a code-based collapse criterion, SLV is evaluated as a function of the ultimate capacity of primary components (first element to reach its displacement or strength capacity in the whole structure). For RC columns, ductile failure allows to exploit the ultimate rotation capacity of the elements. However, the displacement capacity of existing members is often limited by lack of proper reinforcing details. Indeed, in existing buildings designed prior to the introduction of seismic-oriented design codes, structural capacity could be significantly limited due to structural deficiencies such as inadequate detailing of reinforcement (e.g., light transvers reinforcement in columns and no transverse reinforcement in the beam-column joint), deficiencies in anchorage, smooth rebars, and the absence of any capacity design principles [40,41]. Thus, to properly assess the lateral capacity of existing buildings, possible brittle failures due to reduced shear capacity for the columns or in unconfined lateral joints have to suitably identified. Suitable identification of the rotation corresponding to the attainment of flexural or shear capacity allows to define the type of failure for the element. In particular, the rotation at SLV limit state θ_{SLV} is defined with a procedure that is based on the comparison between the yielding shear (V_y), calculated as the ratio between M_y and the shear span of the column (L_v), and the shear strength (V_n) calculated according to [42]. L_v is assumed equal to half of the column height, consistently with the shear type assumption. Depending on the ratio V_y/V_n, three possible failure modes are expected: flexure, flexure-shear or pure shear failure. If $V_y/V_n < 1$ for any value of ductility demand, RC column is expected to fail in flexure and θ_{SLV} corresponds to $\frac{3}{4}$ of the ultimate chord rotation for ductile members [39]. If $V_y/V_n \geq 1$, the column is expected to fail in shear or flexure-shear and θ_{SLV} is evaluated correspondingly to the intersection point between column shear demand and column shear capacity (see Figure 2b).

Beam-column joints of existing GLD buildings typically have no transverse reinforcement in the joint region. Due to lack of shear reinforcement, the shear strength of the joint panel zone is only provided by concrete resistance and can be directly related to the principal tensile strength of the concrete. In this case, using the Mohr's circle for state of stress, the shear strength of joints can be related to principle tensile strength of concrete and to column axial load as follows:

$$V_{jt} = A_j \sqrt{\sigma_{nt,lim}^2 + \sigma_{nt,lim}\frac{N_i}{A_j}} \tag{1}$$

where V_{jt} is the joint tensile strength, $\sigma_{nt,lim}$ is the principal tensile strength, N is the column axial load in the upper column and A_j is the joint cross-sectional area.

In the same way, beam-column joints may potentially fail in shear due to high principal compression stresses. Such failures can be quantified using the principal compression strength instead of the principal tension strength as follows:

$$V_{jc} = A_j \sqrt{\sigma_{nc,lim}^2 - \sigma_{nc,lim} \frac{N_i}{A_j}} \qquad (2)$$

where V_{jc} is the joint compressive strength and $\sigma_{ct,lim}$ is the principal compressive strength.

According to the Italian seismic code [39], the brittle failure of unconfined joints is attained for a limit value of the tensile principal stress of $\sigma_{nt,lim} = 0.3\sqrt{f_c}$ or for a compressive principal stress of $\sigma_{nc,lim} = 0.5f_c$, with f_c expressed in MPa. Adopting the simplified free-body diagram for an external beam-column joint depicted in Figure 1 (lower-right panel), the shear force demand acting at the core of the joint (V_j) can be calculated as $V_j = V_{c,t} - T$, where $V_{c,t}$ is the shear demand of the upper column (subscript t stands for top) and T represents the tension force developing in the steel reinforcement of the beam. It is assumed that $T = M_b/(0.9 \cdot d)$, where M_b is the beam moment and d is the effective depth of beam cross-section ($0.9 \cdot d$ is the internal lever arm). The maximum value of T is limited by yielding strength of longitudinal reinforcement, thus $T \leq A_s \cdot f_y$, where A_s is the area of the beam top reinforcement, and f_y is the yield strength of reinforcing steel. Beam moment, M_b, is calculated based on equilibrium considerations, assuming that the inflection point is located at the mid-height of the columns. In particular, $M_b = V_{c,t} \cdot L_{v,a} + V_{c,b} \cdot L_{v,b}$ where L_v indicates the shear span, subscript a and b indicates the member above and below the considered joint, respectively; and subscript c indicates the column.

The presence of masonry infills can provide a significant contribution to the global response framed structures under seismic loads, increasing both stiffness and lateral strength of RC structures. On the other hand, the presence of infills induces additional shear forces at the ends of beams and columns that may lead to the activation of brittle collapse mechanisms especially in non-ductile RC structures (e.g., Figure 4). The quantification of these additional shear forces was addressed by different authors [43,44] and was found to be depending on both geometrical and mechanical characteristics of the panel. In this work, according to [44] it is assumed that the additional shear force induced in the surrounding columns is equal approximately to the 25% of the horizontal component of axial load experienced by the equivalent strut. Considering this additional shear force acting on the node, the previous formulation is modified in $V_j = V_{c,t} + \gamma V_i - T$, where V_i is the shear force acting in the masonry infill panel and γ is a reduction coefficient that accounts for the proportion of the force that is transferred from the infill to the column (here assumed as $\gamma = 0.25$). Note that in corner columns it often happens that the absolute value of T significantly exceeds the absolute value of $V_{c,t}$; hence, if the effect of the infills is neglected it generally happens that a higher value of V_j is obtained, that is a conservative assumption. On the other hand, due to the additional shear force γV_i on the column, column shear failure may occur at lower drifts with respect to the case where the infill action is neglected.

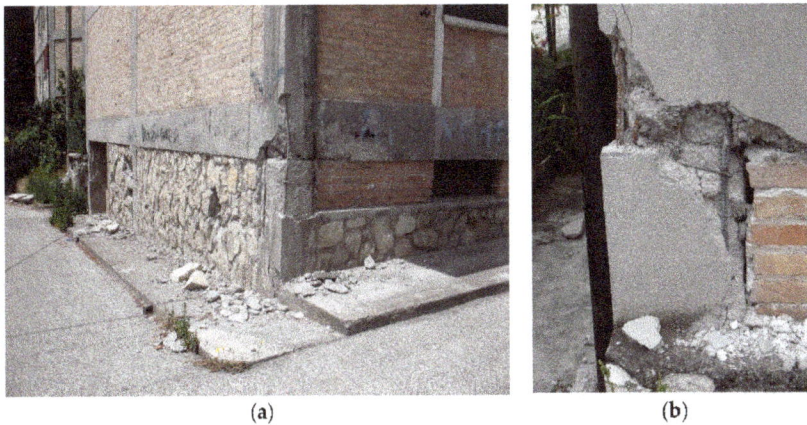

(a)   (b)

Figure 4. Interaction between infills and RC structural members. (**a**) Damage to column and joint, (**b**) detail.

2.3. Seismic Capacity

The seismic capacity, corresponding to the attainment of maximum interstorey drift equal to 2‰ for SLD limit state and to brittle or ductile failure of an RC element for SLV limit state, is detected along the pushover curve. In order to evaluate the building behavior with respect to code prescriptions it is useful to express the capacity in terms of maximum peak ground acceleration PGA_c that the building is able to withstand for each limit state ($PGA_{c,SLD}$ and $PGA_{c,SLV}$). Indeed, the ratio of PGA_c versus the seismic demand expressed by the design peak ground acceleration at the relevant limit state PGA_d, represent the building safety index IS. The building safety index is a widely used parameter to evaluate the performance of a construction with respect to new building standards [45–48].

Starting from the pushover curve, the evaluation of inelastic spectral displacement corresponding to the seismic capacity, and the corresponding PGA_c, can be performed via the capacity spectrum method (CSM) in the modified version proposed in [49,50] for infilled frames.

Following the approach suggested in [50] the pushover curve can be approximated with a quadri-linear curve adopting an equivalent energy principle. In particular, essential parameters describing the quadrilinear are the maximum and minimum strength, F_{max} and F_{min}, the displacement corresponding to the yielding of the equivalent system D_y and the initial and final points of the softening tract D_s and D_m, as shown in Figure 5a. Those data, together with the seismic mass m_i at the generic ith storey and the normalized deformed shape Φ, allow determining the transformation factor Γ and the equivalent single degree of freedom SDOF system, defined by the mass $m^* = \Sigma m_i \Phi_i$, the maximum strength C_{bmax} and the period T^*:

$$C_{bmax} = \frac{F_{max}}{\Gamma m^* g} \tag{3}$$

$$T^* = 2\pi \sqrt{\frac{m^* D_y}{F_{max}}} \tag{4}$$

Two possible shape of vector Φ, representing normalized deformed shape, are assumed, the former proportional to first mode shape and latter proportional to inertia forces induced by constant acceleration along the height—the so-called mass proportional load.

The pushover expressed in terms of the displacement of the equivalent SDOF $\Delta^* = \Gamma \cdot \Delta_{top}$ and C_b is the so-called capacity curve, that can be used to represent the behavior of a SDOF system in a spectral approach, i.e., in the acceleration displacement response spectrum ADRS format. Starting from the defined quadri-linear curve some of the parameters to determine the seismic demand with a spectral approach may be derived. Adopting the R-μ-T relationship introduced in [49] for infilled buildings, in fact, in addition to the elastic period T*, it is necessary to determine also a strength factor r_u, defined as the ratio of minimum strength F_{min} versus maximum one F_{max}, and the ductility μ_s at the beginning of the softening branch, given by the ratio of Δ_s versus Δ_y.

Hence, given the input elastic spectrum, the elastic demand in terms of displacement and acceleration (S_{de}, S_{ae}) is determined in correspondence of the elastic period T* of the SDOF equivalent system; the reduction factor R_μ corresponds to S_{ae}/C_{bmax}. The ductility demand μ is evaluated starting from the R-μ-T relationship introduced in [49]; finally, the inelastic displacement demand S_{di} is obtained multiplying the μ for the displacement at the elastic limit Δ^*_y of the equivalent SDOF. Figure 5b illustrates the application of the CSM for an infilled building. By scaling the elastic spectrum, the corresponding inelastic spectrum is also scaled, and it is possible to find the elastic spectrum such that the inelastic displacement demand corresponds to the spectral seismic capacity of the building. The corresponding anchoring peak ground acceleration is PGA_c.

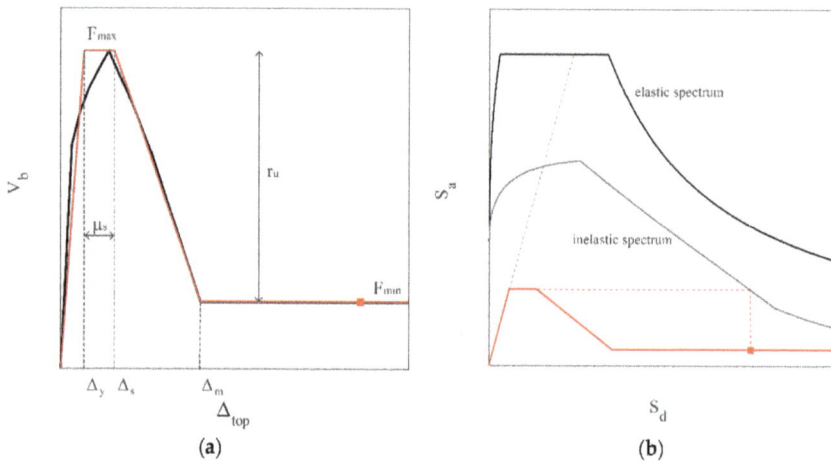

Figure 5. Pushover curve and evaluation of lateral seismic capacity with spectral approach (a) Pushover curve and the quadrilinear idealization according to [50]; and (b) the application of capacity spectrum method for evaluation of the capacity.

In this paper Eurocode 8 spectral shape [20] for a subsoil B category is adopted for exemplification purposes, but other spectral shapes could be equally used.

2.4. Local Retrofit of Joints and Columns with FRP Wrapping

Different strategies can be adopted to enhance the structural performances of existing buildings. The structural capacity may be increased in terms of ductility, stiffness or strength or by reducing the seismic demand adopting many local or global strategies [21,51,52]. Among possible techniques, an effective strategy for existing RC structures could be based on local modification of components that are inadequate in terms of strength or deformation capacity. It aims at increasing the strength or deformation capacity of deficient components to significantly increase the structural seismic capacity even under limited budget constraint. This local strengthening technique has the advantage to require only the assessment of local components capacity increase, without the need to perform a new global analysis of building response, provided that global mass and structural stiffness are not significantly affected by the local strengthening intervention. In this paper, with the aim of investigating the effect of widely used retrofit solutions, the strategy employing externally bonded FRP reinforcement is applied for building upgrading. In particular, the FRP strengthening on corner beam–column joints and the increase of the shear strength of columns with full-height FRP reinforcement are considered. Confinement by FRP of columns and joints results in an increase of the building deformation capacity avoiding brittle failure modes [53].

For what concerns the joint strengthening, the model proposed in [54] is adopted. In such a model it is supposed that the FRP system contributes to the principle tensile stress with a component that depends on the inclination of the fibers, and that the FRP fibers provide a similar contribution as the internal steel reinforcement. To evaluate the global strengthening effect on the corner joints, it is supposed to apply quadriaxial carbon fiber reinforced polymers CFRP fabric with fibers inclined of 0°, ±45°, and 90° at the beam axes. One or more layers of CFRP can be applied, depending on the strength increase to realize. In order to prevent shear failure at the column joint interface due to local effects of infills, steel reinforced polymer SRP composites in the form of uniaxial systems are disposed around the beam column–joint prior to application of CFRP quadriaxial fabric (Figure 6a,b) [55].

The increase of columns shear strength can be obtained with external FRP reinforcement, strengthening the columns with discontinuous or continuous uniaxial CFRP strips, with fibers perpendicular to the column longitudinal axis (see e.g., Figure 6c). Depending on the strength increase to realize, one or more plies of CFRP can be applied. To evaluate the shear capacity of RC columns strengthened with FRP, the FRP contribution can be added to the original capacity of the member, as suggested by international codes provisions [56,57]. The obtained strength increase can be easily taken into account for assessment of the SLV limit state in retrofitted structures [58].

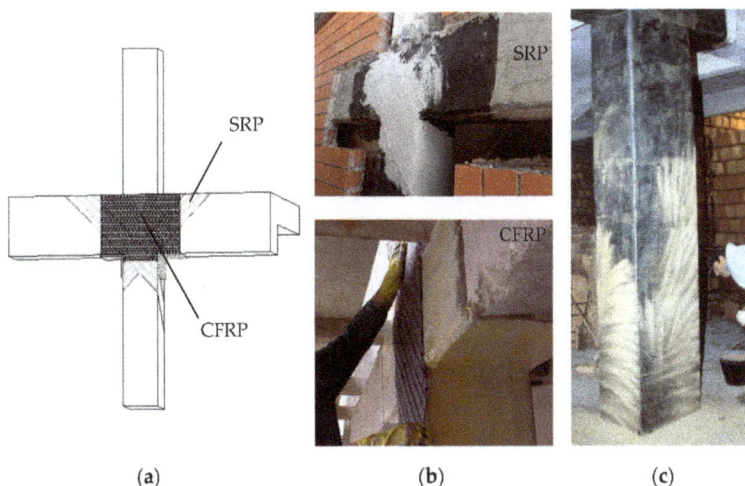

Figure 6. Local retrofit solutions for columns and joints: (a) application of SRP uniaxial system to withstand horizontal action due to infills and quadriaxial CFRP fabric for corner joint (b); and shear strengthening of the columns with uniaxial CFRP strips (c).

3. Example Application for RC Infilled Building Typologies

Adopting the methodology described in Section 2, the model for three archetype regular buildings of rectangular shape, representative of existing GLD RC frames of 3, 5, and 7 storeys are obtained with simulated design. The structural model for each archetype building is fixed, with bay lengths in longitudinal and transversal direction assumed to be $a_x = a_y = 4.3$ m and interstorey height $a_z = 3$ m (see Figure 7). For the three-storey buildings, transversal frame columns dimensions at the first storey are 30×30 cm with column reinforcement is 4φ12 for each column. For the five-storey buildings, transversal frame columns dimensions at the first storey are 30×30 cm for corner columns and 30×40 for interior columns; longitudinal reinforcement consists of 4φ12 for corner columns and of 6φ12 for interior columns. For the seven-storey buildings, transversal frame columns dimensions at the first storey are the same that for the five-storey buildings while the longitudinal reinforcement consists of 6ϕ12 for corner columns and of 8ϕ12 for interior columns. Referring to the concrete and steel properties, a compressive concrete stress f_c = 25 MPa and a steel tensile yielding stress f_y = 399 MPa are chosen as representative values for GLD buildings constructed in the decade '62–'71.

The infills consistency is assumed to be variable depending on the infill thickness and strength. In particular, we assume that weak (W) and medium (M) panels are realized with a double layer brick infill having (80 + 80) mm or (120 + 120) mm thickness, respectively (global thickness t_w = 160 mm for W and t_w = 240 mm for M), while a single layer brick infill of (300) mm thickness is assumed for strong (S) panels; these infill masonry configurations are widely used in European building practice [59]. Concerning the elastic shear modulus, it is assumed a mean value G_w = 1089 MPa for W and M infills and G_w = 1296 MPa for S ones. These values are chosen according to the proposal of [59] and are deemed to represent increasing infills stiffness. The E_w is assumed equal to 10/3 G_w coherently to the ratio that can be inferred from [26], while cracking strength of the masonry is considered linearly dependent on G_w according to boundary values indicated in [26].

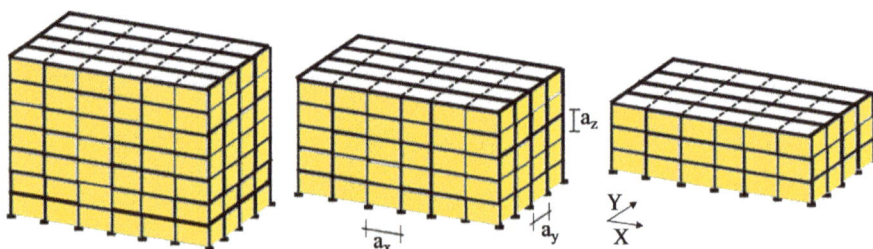

Figure 7. Scheme of the archetype buildings considered for the analyses.

3.1. Lateral Seismic Capacity for As-Built and Retrofitted Configurations

Each building model is analyzed with simplified pushover analysis in both the longitudinal and the transversal direction and considering mode and mass proportional lateral load distribution. Figure 8 shows the capacity curves obtained in transversal direction for mass proportional loads for 3 (a), 5 (b) and 7 (c) storey buildings and considering W (black line), M (thin gray line) and S infills (thick gray line). SLD and SLV limit states are plotted as yellow and red dots along the curves. Only the transversal direction is represented because it corresponds to the lower capacity in terms of PGA. Apparently, the SLD limit state is not significantly influenced by the infill type. However, the $PGA_{c,SLD}$ is higher for stronger infills; for example $PGA_{c,SLD}$ increases from 0.06 g to 0.11 g from 3W to 3S or from 0.04 g to 0.06 g from 7W to 7S. This happens due to the increased stiffening effect for S configurations, that leads to lower T* and to a consequently higher PGA capacity. Additionally, from Figure 8 it is noted that the attainment of SLV limit state is premature and occurs before the SLD limit state; this is due to brittle failure of columns or of exterior unconfined nodes (see Figure 4) according to the code criteria [39]. This circumstance is common for existing buildings designed only for gravity loads and not respecting capacity design principles [55]. On the other hand, these types of brittle failures can be avoided with application of local retrofit interventions. In this study we simulate two possible type of retrofit solutions. The first one, named FRP-only, corresponds to the application of quadriaxial CFRP fabric for unconfined lateral joints as well as strengthening of the columns with continuous uniaxial CFRP strips; three increasing levels of retrofit are considered, namely R1, R2 and R3, corresponding to the application of 1, 2, or 3 plies of FRP for the elements (i.e., exterior beam column joints or columns) that are not verified at the SLV limit state. Second type, named FRP + SRP, that is analogous to the first one, but in addition to FRP, also SRP uniaxial system is disposed around the beam column–joint prior to application of CFRP quadriaxial fabric. Both types of retrofit can be modeled as described in Section 2.4.

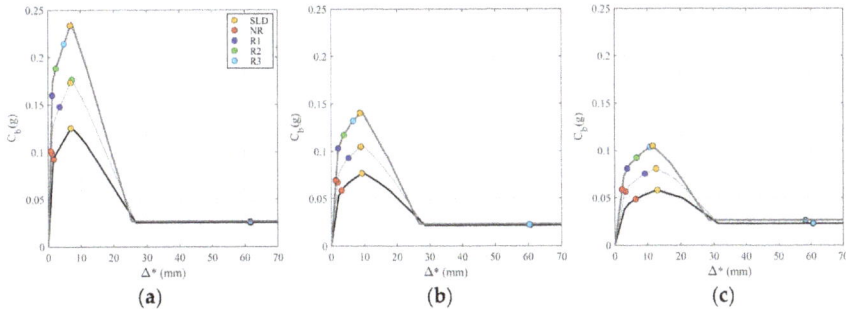

Figure 8. Capacity curves for the archetype buildings in transversal direction with mass proportional loads: (**a**) three-storey, (**b**) five-storey, and (**c**) seven-storey building with W (black), M (gray thin) and S (gray thick) infills. Limit states for no retrofit NR and increasing retrofit levels of FRP-only solution are represented along the curves.

Table 1 resumes the results of the pushover analyses with mass and mode proportional horizontal forces and the capacity at SLV limit state for all the considered cases, i.e., for buildings of 3, 5, and 7 storeys with W, M, and S infills configuration and considering the no retrofit NR as well as R1, R2, and R3 retrofit levels for FRP-only and FRP + SRP retrofit solutions. Note that SRP is designed to avoid infill-induced failure and no different retrofit levels are considered as in the case of FRP (i.e., R1, R2, and R3). The mechanism type, when ductile behavior is fully exploited, is always local mechanism at the first storey, except for the case of mode proportional forces and 7W configuration, where the soft storey is at the second level. The table indicates the $PGA_{c,SLV}$ (g) that is attained for each case. Note that the maximum value of $PGA_{c,SLV}$ (g) for each building configuration corresponds to a ductile type mechanism (indicated as $PGA_{c,ductile}$ and reported in bold in Table 1). The table displays also the type of failure for each case, indicated as nS-C and nS-J for column failure (brittle or ductile) or joint failure at nth storey, respectively. If the $PGA_{c,SLV}$ < $PGA_{c,ductile}$ the failure is brittle, either for joints or for columns in shear. Applying either the first or the second type of retrofit the attainment of SLV limit state is significantly delayed and the lateral seismic capacity increases. However, the effect is sensibly different, especially for the configurations with strong S infills. Blue, green, and cyan dots along the pushover curves in Figure 8 represent the varied SLV limit states corresponding to R1, R2, and R3 of FRP-only solution. As it can be seen from the figures and Table 1, for the case of W infills, it is sufficient the R1 level of FRP-only solution (except for 3W for which R2 level is needed) to allow developing of a ductile type mechanism (in this case blue, green, and cyan dots are superimposed in Figure). On the other hand, for S infills, even increasing significantly the retrofit level (e.g., from R1 to R3), the seismic capacity corresponding to the formation of a ductile type mechanism is not reached. For the weak W configuration $PGA_{c,SLV}$ varies from 0.022 g to 0.15 g for the 3W configuration, from 0.017 g to 0.15 g for 5W, and from 0.026 g to 0.14 g for 7W with the R1 level. On the other hand, for the S configuration, even the R3 retrofit level does not allow the structure to reach the displacement capacity on the plateau of the pushover curve: correspondingly, $PGA_{c,SLV}$ varies from 0.02 g to 0.03 g, 0.06 and 0.09 for the 3S configuration, and for R1, R2, and R3 levels, from 0.016 g to 0.024 g, 0.043 and 0.06 for 5S R1, R2, and R3, and from 0.014 g to 0.024 g, 0.04 g, and 0.055 g for 7S R1, R2, and R3, respectively.

Table 1. PGA$_{c,SLV}$ (g) for the archetype buildings with W, M and S infills configurations and for mass/mode analyses. No retrofit NR and R1, R2 R3 increasing retrofit levels are considered for FRP-only and FRP + SRP retrofit solutions. nS-C and nS-J indicates column failure (brittle or ductile) or joint failure at the nth storey, respectively.

Model	NR	FRP-Only			FRP + SRP		
		R1	R2	R3	R1	R2	R3
3W	0.022/0.024 1S-C/1S-C	0.022/0.024 1S-C/1S-C	0.151/0.141 1S-C/1S-C	0.151/0.141 1S-C/1S-C	0.151/0.077 1S-C/1S-J	0.151/0.141 1S-C/1S-C	0.151/0.141 1S-C/1S-C
3M	0.02/0.02 1S-C/1S-C	0.060/0.065 1S-C/1S-C	0.085/0.094 1S-C/1S-C	0.145/0.140 1S-C/1S-C	0.145/0.140 1S-C/1S-C	0.145/0.140 1S-C/1S-C	0.145/0.140 1S-C/1S-C
3S	0.020/0.022 1S-C/1S-C	0.032/0.035 1S-C/1S-C	0.060/0.064 1S-C/1S-C	0.093/0.098 1S-C/1S-C	0.147/0.148 1S-C/1S-C	0.147/0.148 1S-C/1S-C	0.147/0.148 1S-C/1S-C
5W	0.017/0.018 1S-C/1S-C	0.148/0.136 1S-C/1S-C	0.148/0.136 1S-C/1S-C	0.148/0.136 1S-C/1S-C	0.055/0.136 1S-J/1S-C	0.148/0.136 1S-C/1S-C	0.148/0.136 1S-C/1S-C
5M	0.015/0.016 1S-C/1S-C	0.038/0.044 1S-C/1S-C	0.146/0.137 1S-C/1S-C	0.146/0.137 1S-C/1S-C	0.146/0.137 1S-C/1S-C	0.146/0.137 1S-C/1S-C	0.146/0.137 1S-C/1S-C
5S	0.015/0.016 1S-C/1S-C	0.024/0.025 1S-C/1S-C	0.043/0.047 1S-C/1S-C	0.060/0.070 1S-C/1S-C	0.142/0.095 1S-C/1S-J	0.142/0.138 1S-C/1S-C	0.142/0.138 1S-C/1S-C
7W	0.026/0.020 1S-C/2S-C	0.139/0.125 1S-C/2S-C	0.139/0.125 1S-C/2S-C	0.139/0.125 1S-C/2S-C	0.139/0.125 1S-C/2S-C	0.139/0.125 1S-C/2S-C	0.139/0.125 1S-C/2S-C
7M	0.016/0.014 1S-C/2S-C	0.040/0.044 1S-C/2S-C	0.145/0.138 1S-C/1S-C	0.145/0.138 1S-C/1S-C	0.145/0.138 1S-C/1S-C	0.145/0.138 1S-C/1S-C	0.145/0.138 1S-C/1S-C
7S	0.014/0.014 1S-C/3S-C	0.024/0.026 1S-C/1S-C	0.039/0.047 1S-C/1S-C	0.055/0.070 1S-C/1S-C	0.146/0.142 1S-C/1S-C	0.146/0.142 1S-C/1S-C	0.146/0.142 1S-C/1S-C

Figure 9a schematically represents the effect of the FRP-only retrofit for the case of 5W and 5S configurations. C shear (J shear) in figure indicates brittle shear failure in column (corner joint) and C flex indicates ductile failure in the column.

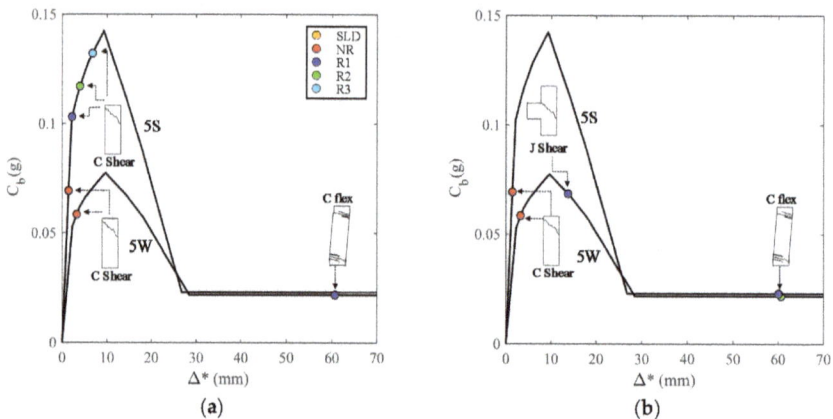

Figure 9. Pushover curves for 5W and 5S configurations and failure mechanisms for the NR and (a) R1, R2, and R3 FRP-only retrofit solutions and (b) R1, R2, and R3 FRP + SRP retrofit solutions.

Analogously to Figure 9a,b schematically represents the effect of the FRP + SRP retrofit for the case of 5W and 5S configurations. In this case, differently from FRP-only solution, the retrofit level R1 allows the change of failure type and with level R2 retrofit (R1 for strong infills) the full lateral seismic capacity, corresponding to a ductile mechanism, is exploited.

The scarce increase of PGA$_{c,SLV}$ for the case of strong infills is due to the significant increment of shear load on columns that is transferred from the adjacent strong infills to the columns and to the consequent brittle failure in shear of the columns. On the other hand, when FRP + SRP strengthening solution is applied, the negative effect of the horizontal action transferred from the infills to adjacent columns is avoided and, even in the case of S infills, it is generally sufficient the retrofit level R1 to attain full development of the ductile type mechanism. Figure 10a–c show the same capacity curves as in Figure 8, but the SLV limit states corresponding to R1, R2, and R3 of FRP + SRP solution are displayed along the curves (instead of FRP-only as in Figure 8).

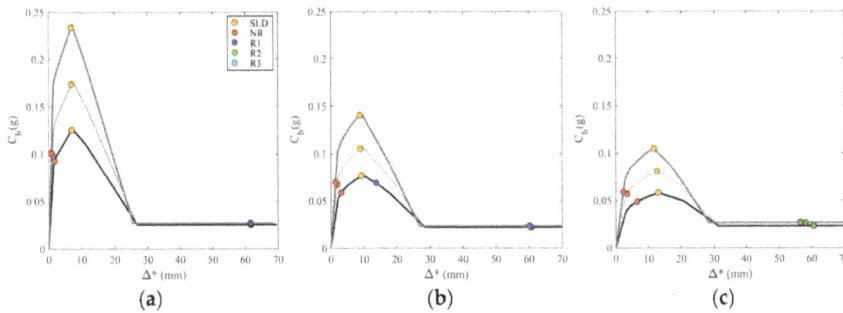

Figure 10. Capacity curves for the archetype buildings in transversal direction with mass proportional loads: (**a**) three-storey (**b**) five-storey, and (**c**) seven-storey building with W (bold black), M (thin gray) and S (bold gray) infills. Limit states for no retrofit NR and increasing retrofit levels of FRP + SRP solution are represented along the curves.

As it can be seen, when the transferring of horizontal action from the infills to adjacent columns is prevented, by application of SRP reinforcement, for most of the considered cases it is sufficient the retrofit level R1 to allow exploiting a ductile type mechanism. This is confirmed by the increase of $PGA_{c,SLV}$ that is attained even by the application of the R1 retrofit level for either the W and the S configurations. Indeed, $PGA_{c,SLV}$ varies from 0.022 g to 0.15 g for the 3W configuration R1 level, from 0.017 g to 0.055 g for 5W R1, to 0.15 for 5W R2, and from 0.026 g to 0.14 g for 7W with the R1 level. For the case of FRP + SRP similar results are obtained for the S configuration: $PGA_{c,SLV}$ varies from 0.020 g to 0.15 g for the 3S configuration R1 level, from 0.016 g to 0.14 g for 5S R1, and from 0.014 g to 0.15 g for 7S with the R1 level.

3.2. Sensitivity Analysis

Considering the archetype buildings and models introduced in the previous section, a sensitivity analysis is carried out introducing possible variability of infill consistency. In particular, a Monte Carlo simulation is performed for each considered infill configuration, W, M and S, where the t_w is not varied with respect to previous assumptions, while G_w is varied assuming a lognormal distribution with median values indicated in Section 3 and considering a COV = 30% for each building. For each of the obtained configurations the $PGA_{c,SLD}$, and $PGA_{c,SLV}$, are calculated as described in Section 2.3.

Figure 11 shows the variation of $PGA_{c,SLD}$ with $G_w t_w$. As observed in Section 3.1, the seismic capacity at the SLD limit state tends to increase from weaker infills (represented by lower $G_w t_w$) to stronger ones. This is mainly due to the increase in lateral stiffness and to the consequent lowering of equivalent period T^* and related seismic drift demand. For the same reason, the seismic capacity at SLD for higher buildings, that are characterized by higher T^* with respect to shorter buildings, is generally lower.

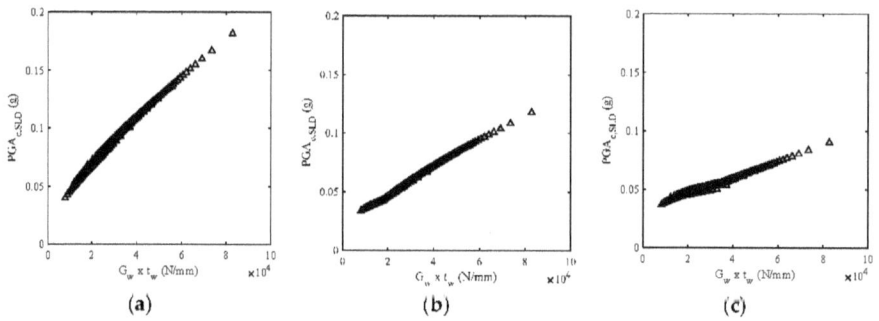

Figure 11. Variation of PGA$_{c,SLD}$ with G$_w$t$_w$ for (**a**) three-storey, (**b**) five-storey, and (**c**) seven-storey buildings.

Figure 12 shows the variation of PGA$_{c,SLV}$ with G$_w$t$_w$ comparing the cases of no retrofit NR and increasing levels of retrofit solution FRP-only (R1 and R2). On ordinates the ratio of PGA$_{c,SLV}$ versus the maximum value that can be reached for the same configuration for a ductile mechanism, PGA$_{c,ductile}$, is represented.

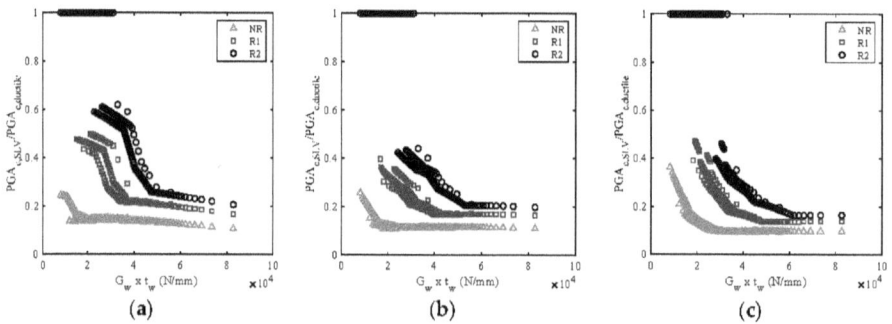

Figure 12. Variation of PGA$_{c,SLV}$/PGA$_{c,ductile}$ with G$_w$t$_w$ for (**a**) three-storey, (**b**) five-storey, and (**c**) seven-storey buildings. The effect of retrofit solution FRP-only is also represented.

A decreasing trend of the ratio PGA$_{c,SLV}$/PGA$_{c,ductile}$ with increasing G$_w$t$_w$ is observed for either the NR and the retrofitted solutions. As a general comment it is noted that for weaker infills, characterized by lower values of the product G$_w$t$_w$, it is possible a significant enhancement of the seismic capacity with the application of retrofit level R1 or R2 of the FRP-only solution. On the other hand, with increasing infills stiffness and strength, the benefit of seismic strengthening with only FRP is sensibly reduced, with just a minimum capacity increase for stronger infills.

Figure 13 shows the variation of PGA$_{c,SLV}$/PGA$_{c,ductile}$ with G$_w$t$_w$ comparing the cases of no retrofit NR and increasing levels of retrofit solution FRP + SRP (R1 and R2). Differently from previous case, where the presence of increasing level of horizontal actions transferred from infills to adjacent columns and the consequent brittle failure of columns penalizes the capacity of stronger infills configurations, here the PGA$_{c,SLV}$ for the retrofitted solutions increases with G$_w$t$_w$. In particular, for most of the R1 cases in three-storey and five-storey buildings, it is observed an increasing capacity growth with G$_w$t$_w$ and for seven-storey building the R1 retrofit level allows to reach PGA$_{c,ductile}$ even for weaker infills (lower G$_w$t$_w$) configurations. In some cases, not evident in figure because the points are overlapped with R2 retrofit level, it can happen that R1 is enough to reach PGA$_{c,ductile}$ even for the three- and five-storey buildings with lower G$_w$t$_w$.

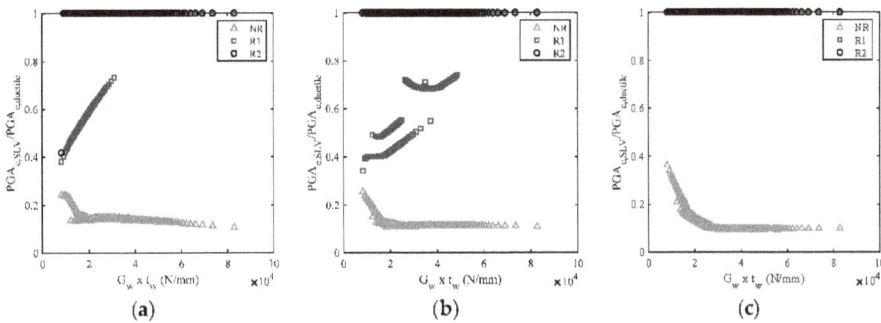

Figure 13. Variation of $PGA_{c,SLV}$/ $PGA_{c,ductile}$ with $G_w t_w$ for (**a**) three-storey, (**b**) five-storey, and (**c**) seven-storey building. The effect of retrofit solution FRP + SRP is also represented.

4. Discussion

An investigation on the effect of infills on the lateral seismic capacity of existing gravity load designed GLD reinforced concrete RC infilled frames was performed. Possible brittle failures in unconfined lateral joints or in columns, that are a common cause of collapse for existing RC buildings not designed for seismic loads or with obsolete seismic provisions, were explicitly considered for the evaluation of seismic capacity. Archetype buildings representative of existing gravity load designed RC frames of three different height ranges (three-, five-, and seven-storey) were obtained with a simulated design process and studied via simplified pushover analyses, including the effect of the infills and considering frame-infill interaction. Three different infills consistency are considered in the analyses, namely weak, medium and strong. The information on infill consistency is generally missing in large scale vulnerability studies. However, this information is collected in the interview-based Cartis form [60], recently implemented by Italian Civil Protection.

Moreover, possible alternative local retrofit interventions devoted to avoiding brittle failures were considered, evaluating their relative efficacy in case of weak or stronger infills. Additionally, a sensitivity analysis was performed to assess the influence of infills consistency variation on the lateral seismic capacity at damage limitation and life safety limit states. Considering the budget constraints that typically affect seismic mitigation campaigns for seismic vulnerable building typologies, which are widespread in Italy, local retrofit interventions are one of the preferred solutions, allowing upgrading of the seismic performance of the buildings as well as the containment of the retrofit costs. Therefore, the results of this paper can give useful preliminary indication on the level of local retrofit solutions required depending on if the infills are strong or weak. Further experimental studies would be beneficial to support the analytical findings from this study.

5. Conclusions

The study evidenced that for the archetype analyzed existing GLD buildings, designed in absence of seismic provisions and of capacity design rules, the attainment of SLV limit state is premature and occurs before the SLD limit state; this is due to brittle failure of columns or of exterior unconfined nodes. The lateral seismic capacity at SLV limit state can be increased by the application of local retrofit interventions, allowing to avoid the brittle failure types for the upgraded elements. However, the efficacy of the retrofit intervention varies depending on the consistency of infills.

If FRP-only retrofit solution is applied in critical elements or joints (FRP strengthening of unconfined lateral joints as well as of the columns in shear), it is noted that as the infills consistency increases the benefit of the retrofit reduces, and in general for stronger infills just a minimum capacity enhancement is attained even adopting increasing level, and costs, of retrofit. For weaker infills a higher benefit in terms of seismic capacity is observed, although it is generally not possible to reach the capacity corresponding to the full development of a ductile type mechanism, even with higher levels of retrofit. This trend mainly depends on the effect of the concentrated horizontal action transferred from the infills to the surrounding frame elements: the shear increase, that is higher for stronger infills, determines a premature shear brittle failure in the columns that can be hardly avoided even with application of a high level of retrofit (e.g., R3 in the example). For weaker infills the concentrated lateral action is lower, and it is easier to prevent brittle failures with application of local retrofit interventions.

However, a significantly higher benefit can be obtained adopting the FRP + SRP retrofit solution, where in addition to FRP strengthening of unconfined lateral joints and of the columns in shear, steel reinforced polymer SRP strips are applied to withstand horizontal action due to infills. In this case, it is observed that the capacity increases as the infills consistency grows for the case of three and five storeys and considering a primary level of building retrofit (R1 in the example), while for a higher retrofit level (R2 in the example) it is always possible to attain the lateral seismic capacity corresponding to a ductile type mechanism. In fact, the application of a specific local reinforcement, the SRP strips, allows to absorb the high concentrated action that is transferred from the infills and the full exploitation of the benefic effect of the FRP retrofitting of joints and columns in shear. When the aggravating effect of concentrated shear due to frame-infill interaction is mitigated (e.g., through SRP strips or similar solution based on the same principle) the infills consistency does not limit anymore the full exploitation of lateral seismic capacity. On the contrary, the application of a relatively low level of retrofit (R1 in the example), allows obtaining an increasing level of capacity enhancement for stronger infills.

Author Contributions: M.P., M.G.d.A, and M.D.L. collected scientific articles for state of the art. M.G.d.A. performed the analyses, analyzed the results, and contributed to writing the paper. M.P. organized the work, contributed to the analyses of the results, and to writing the paper. M.D.L. gave advice on the alternative retrofit strategies and contributed to the analyses of results. A.P. contributed to the analyses of results and revised the paper for overall consistency and scopes.

Acknowledgments: This study was performed in the framework of PE 2018; joint program DPC-Reluis Subproject RC-WP1: vulnerability of RC construction at the territorial scale and Subproject TT: Territorial themes.

Conflicts of Interest: The authors declare no conflict of interest.

References

1. Grunthal, G. (Ed.) *Chaiers du Centre Européen de Géodynamique et de Séismologie: Volume 15—European Macroseismic Scale 1998*; European Center for Geodynamics and Seismology: Luxembourg, 1998.
2. Rojan, C. *ATC-20-1 Field Manual: Postearthquake Safety Evaluation of Buildings*; Applied Technology Council: Redwood City, CA, USA, 2005.
3. Polese, M.; Gaetani d'Aragona, M.; Di Ludovico, M.; Prota, A. Sustainable selective mitigation interventions towards effective earthquake risk reduction at the community scale. *Sustainability* **2018**, *10*, 2894. [CrossRef]
4. Bazzurro, P.; Mollaioli, F.; De Sortis, A.; Bruno, S. Effects of masonry walls on the seismic risk of reinforced concrete frame buildings. In Proceedings of the 8th US National Conference on Earthquake Engineering and Seismology, San Francisco, CA, USA, 18–22 April 2006.
5. Borzi, B.; Crowley, H.; Pinho, R. The influence of infill panels on vulnerability curves for RC buildings. In Proceedings of the 14th World Conference on Earthquake Engineering, Beijing, China, 12–17 October 2008.
6. Negro, P.; Colombo, A. Irregularities induces by nonstructural masonry panels in framed buildings. *Eng. Struct.* **1997**, *19*, 576–585. [CrossRef]
7. Dolšek, M.; Fajfar, P. Soft storey effects in uniformly infilled reinforced concrete frames. *J. Earthq. Eng.* **2001**, *5*, 1–12. [CrossRef]
8. Fardis, M.N. Seismic design issues for masonry-infilled RC frames. In Proceedings of the First European Conference on Earthquake Engineering and Seismology, Geneva, Switzerland, 3–8 September 2006.
9. Comité Euro-International du Béton (CEB). *RC Frames under Earthquake Loading: State of the Art Report*; Bulletin 231; Thomas Telford Ltd.: London, UK, 1996.
10. Kakaletsis, D.J.; David, K.N.; Karayannis, C.G. Effectiveness of some conventional seismic retrofitting techniques for bare and infilled R/C frames. *Struct. Eng. Mech.* **2011**, *39*, 499–520. [CrossRef]

11. Calvi, G.M.; Bolognini, D.; Penna, A. Seismic performance of masonry-infilled RC frames: Benefits of slight reinforcement. In Proceedings of the Sismica 2004-6° Congresso Nacional de Sismologia e Engenharia Sismica, 14–15 April 2004; Available online: ftp://ftp.ecn.purdue.edu/spujol/Mason/New%20Folder/253 -276_G_Michele_Calvi.pdf (accessed on 27 September 2018).

12. Akyuz, U.; Yakut, A.; Ozturk, M.S. Effect of masonry infill walls on the lateral behavior of buildings. In Proceedings of the First European Conference on Earthquake Engineering and Seismology, Geneva, Switzerland, 3–8 September 2006.

13. Gaetani d'Aragona, M.; Polese, M.; Prota, M. Influence factors for the assessment of maximum lateral seismic deformations in Italian multistorey RC buildings. In Proceedings of the COMPDYN 2017 6th ECCOMAS Thematic Conference on Computational Methods in Structural Dynamics and Earthquake Engineering, Rhodes Island, Greece, 15–17 June 2017.

14. Gaetani d'Aragona, M.; Polese, M.; Cosenza, E.; Prota, A. Simplified assessment of maximum interstory drift for RC buildings with irregular infills distribution along the height. *Bull. Earthq. Eng.* **2018**. [CrossRef]

15. Colajanni, P.; Impollonia, P.; Paia, M. The effect of infill panels models uncertainties on the design criteria effectiveness for RC frames. In Proceedings of the Final Workshop of Joint DPC-RELUIS Project 2005–2008, Rome, Italy, 29–30 May 2008; pp. 401–408. (In Italian)

16. Polese, M.; Verderame, G.M. Seismic capacity of RC infilled frames: A parametric analysis. In Proceedings of the XIII ANIDIS National Conference "Seismic Engineering in Italy", Bologna, Italy, 28 June–2 July 2009. (In Italian)

17. Dymiotis, C.; Kappos, A.J.; Chryssanthopoulos, M.K. Seismic reliability of masonry-infilled RC frames. *J. Struct. Eng.* **2001**, *127*, 296–305. [CrossRef]

18. Celarec, D.; Ricci, P.; Dolšek, M. The sensitivity of seismic response parameters to the uncertain modelling variables of masonry-infilled reinforced concrete frames. *Eng. Struct.* **2012**, *35*, 165–177. [CrossRef]

19. Perrone, D.; Leone, M.; Aiello, M.A. Non-linear behaviour of masonry infilled RC frames: Influence of masonry mechanical properties. *Eng. Struct.* **2017**, *150*, 875–891. [CrossRef]

20. Comitée Européen de Normalisation. *European Standard EN 1998-1: Eurocode8. Design of Structures for Earthquake Resistance. Part 1. General Rules, Seismic Actions and Rules for Buildings*; CEN: Brussels, Belgium, 2005.

21. Ministerial Decree, D.M. 20.02.2018 "Updating of Technical Standards for Construction". Available online: http://www.gazzettaufficiale.it/eli/gu/2018/02/20/42/so/8/sg/pdf (accessed on 27 September 2018).

22. Fajfar, P. Capacity spectrum method based on inelastic demand spectra. *Earthq. Eng. Struct. Dyn.* **1999**, *28*, 979–993. [CrossRef]

23. Verderame, G.M.; Polese, M.; Mariniello, C.; Manfredi, G. A simulated design procedure for the assessment of seismic capacity of existing reinforced concrete buildings. *Adv. Eng. Softw.* **2010**, *41*, 323–335. [CrossRef]

24. Polese, M.; Marcolini, M.; Zuccaro, G.; Cacace, F. Mechanism Based Assessment of Damaged-Dependent Fragility curves for RC building classes. *Bull. Earthq. Eng.* **2015**, *13*, 1323–1345. [CrossRef]

25. R.D.L. n 2229/1939 Regulations for the Execution of Simple and Reinforced Concrete Constructions. 1939. Available online: http://www.normattiva.it/atto/caricaDettaglioAtto?atto.DataPubblicazioneGazzetta= 1940-04-18&atto.codiceRedazionale=039U2232 (accessed on 27 September 2018). (In Italian)

26. Circolare del Ministero dei Lavori Pubblici n. 1472 del 23/5/1957. Armature delle Strutture in Cemento Armato. 1957. Available online: http://sttan.it/norme/Storiche/1939_11_10_RDL_n_2229_norme_CA.pdf (accessed on 27 September 2018). (In Italian)

27. Polese, M.; Verderame, G.M.; Manfredi, G. Static vulnerability of existing RC buildings in Italy: A case study. *Struct. Eng. Mech.* **2011**, *39*, 599–620.

28. Ricci, P. Seismic Vulnerability of Existing RC Buildings. Ph.D. Thesis, University of Naples Federico II, Naples, Italy, 2018.

29. Mollaioli, F.; Bazzurro, P.; Bruno, S.; De Sortis, A. Influenza della modellazione strutturale sulla risposta sismica di telai in cemento armato tamponati. In Proceedings of the Atti del XIII Convegno ANIDIS "L'ingegneria Sismica in Italia", Bologna, Italy, 28 June–2 July 2009. (In Italian)

30. Biskinis, D.; Fardis, M.N. Deformations at flexural yielding of members with continuous or lap-spliced bars. *Struct. Concr.* **2010**, *11*, 128–138. [CrossRef]

31. Haselton, C.B.; Liel, A.B.; Taylor-Lange, S.; Deierlein, G.G. *Beam-Column Element Model Calibrated for Predicting Flexural Response Leading to Global Collapse of RC Frame Buildings*; PEER Report 2007; Pacific Engineering Research Center, University of California: Berkeley, CA, USA, 2008; p. 3.
32. Noh, N.M.; Liberatore, L.; Mollaioli, F.; Tesfamariam, S. Modelling of masonry infilled RC frames subjected to cyclic loads: State of the art review and modelling with OpenSees. *Eng. Struct.* **2017**, *150*, 599–621.
33. Panagiotakos, T.B.; Fardis, M.N. Seismic response of infilled RC frames structures. In Proceedings of the 11th World Conference on Earthquake Engineering, Acapulco, Mexico, 23–28 June 1996.
34. Fardis, M.N.; Carvalho, E.C.; Fajfar, P.; Pecker, A. *Seismic Design of Concrete Buildings to Eurocode 8*; CRC Press: Boca Raton, FL, USA, 2015.
35. Mainstone, R.J. On the Stiffnesses and Strengths of Infilled Frames. In Proceedings of the Institution of Civil Engineering, Supplement IV, London, UK, 1971; pp. 57–90. Available online: https://copac.jisc.ac.uk/id/38779199?style=html&title=ON%20THE%20STIFFNESS%20AND%20STRENGTHS%20OF%20INFILLED%20FRAMES (accessed on 27 September 2018).
36. Morandi, P.; Hak, S.; Magenes, G. Performance-based interpretation of in-plane cyclic tests on RC frames with strong masonry infills. *Eng. Struct.* **2018**, *156*, 503–521. [CrossRef]
37. Di Trapani, F.; Macaluso, G.; Cavaleri, L.; Papia, M. Masonry infills and RC frames interaction: Literature overview and state of the art of macromodeling approach. *Eur. J. Environ. Civ. Eng.* **2015**, *19*, 1059–1095. [CrossRef]
38. Ricci, P.; Verderame, G.M.; Manfredi, G. Simplified analytical approach to seismic vulnerability assessment of Reinforced Concrete buildings. In Proceedings of the XIV Convegno ANIDIS "L'ingegneria Sismica in Italia", Bari, Italy, 18–22 September 2011.
39. Circolare del Ministero dei Lavori Pubblici n. 617 del 2/2/2009. Istruzioni per L'applicazione delle "Nuove Norme Tecniche per le Costruzioni" di cui al D.M. 14 Gennaio 2008. G.U. n. 47 del 26/2/2009. 2009. Available online: http://www.gazzettaufficiale.it/eli/id/2009/02/26/09A01318/sg (accessed on 27 September 2018). (In Italian)
40. Fabbrocino, G.; Verderame, G.M.; Polese, M. Probabilistic steel stress–crack width relationship in RC frames with smooth rebars. *Eng. Struct.* **2007**, *29*, 1–10. [CrossRef]
41. Gaetani d'Aragona, M.; Polese, M.; Elwood, K.; Baradaran Shoraka, M.; Prota, A. Aftershock collapse fragility curves for non-ductile RC buildings: A scenario-based assessment. *Earthq. Eng. Struct. Dyn.* **2017**, *46*, 2083–2102. [CrossRef]
42. Comitée Européen de Normalisation, European Standard EN 1998-3: Eurocode 8: Design of Structures for Earthquake Resistance, Part 3: Assessment and Retrofitting of Buildings. 2005. Available online: https://www.saiglobal.com/PDFTemp/Previews/OSH/IS/EN/2005/I.S.EN1998-3-2005.pdf (accessed on 27 September 2018).
43. Celarec, D.; Dolšek, M. Practice-oriented probabilistic seismic performance assessment of infilled frames with consideration of shear failure of columns. *Earthq. Eng. Struct. Dyn.* **2013**, *42*, 1339–1360. [CrossRef]
44. Verderame, G.M.; De Luca, F.; Ricci, P.; Manfredi, G. Preliminary analysis of a soft-storey mechanism after the 2009 L'Aquila earthquake. *Earthq. Eng. Struct. Dyn.* **2011**, *40*, 925–944. [CrossRef]
45. *FEMA 308 Repair of Earthquake Damaged Concrete and Masonry Wall Buildings*; Federal Emergency Management Agency: Washington, DC, USA, 1998.
46. Polese, M.; Marcolini, M.; Gaetani d'Aragona, M.; Cosenza, E. Reconstruction policies: Explicitating the link of decisions thresholds to safety level and costs for RC buildings. *Bull. Earthq. Eng.* **2017**, *15*, 759–785. [CrossRef]
47. Polese, M.; Di Ludovico, M.; Marcolini, M.; Prota, A.; Manfredi, G. Assessing reparability: Simple tools for estimation of costs and performance loss of earthquake damaged RC buildings. *Earthq. Eng. Struct. Dyn.* **2015**, *44*, 1539–1557. [CrossRef]
48. Polese, M.; Di Ludovico, M.; Prota, A. Post-earthquake reconstruction: A study on the factors influencing demolition decisions after 2009 L'Aquila earthquake. *Soil Dyn. Earthq. Eng.* **2018**, *105*, 139–149. [CrossRef]
49. Dolšek, M.; Fajfar, P. Inelastic spectra for infilled reinforced concrete frames. *Earthq. Eng. Struct. Dyn.* **2004**, *33*, 1395–1416. [CrossRef]
50. Dolšek, M.; Fajfar, P. Simplified non-linear seismic analysis of infilled reinforced concrete frames. *Earthq. Eng. Struct. Dyn.* **2005**, *34*, 49–66. [CrossRef]

51. Gaetani d'Aragona, M.; Polese, M.; Prota, A. Relationship between the variation of seismic capacity after damaging earthquakes, collapse probability and repair costs: Detailed evaluation for a non-ductile building. In Proceedings of the 15th ECCOMAS Thematic Conference on Computational Methods in Structural Dynamics and Earthquake Engineering, Crete Island, Greece, 25–27 May 2015.

52. Fédération Internationale du Beton (FIB). *Seismic Assessment and Retrofit of Reinforced Concrete Buildings Externally Bonded FRP Reinforcement for RC Structures, Bulletin 24*; FIB: Lausanne, Switzerland, 2003; p. 312.

53. Ilki, A.; Tore, E.; Demir, C.; Comert, M. Seismic Performance of a Full-Scale FRP Retrofitted Sub-standard RC Building. In *Recent Advances in Earthquake Engineering in Europe: 16th European Conference on Earthquake Engineering, Thessaloniki, 2018*; Springer International Publishing: New York, NY, USA, 2018; pp. 519–544.

54. Del Vecchio, C.; Di Ludovico, M.; Prota, A.; Manfredi, G. Analytical model and design approach for FRP strengthening of non-conforming RC corner beam–column joints. *Eng. Struct.* **2015**, *87*, 8–20. [CrossRef]

55. Frascadore, R.; Di Ludovico, M.; Prota, A.; Verderame, G.M.; Manfredi, G.; Dolce, M.; Cosenza, E. Local strengthening of reinforced concrete structures as a strategy for seismic risk mitigation at regional scale. *Earthq. Spectra* **2015**, *31*, 1083–1102. [CrossRef]

56. CNR-DT 200 R1/2013. Guide for the Design and Construction of Externally Bonded FRP Systems for Strengthening Existing Structures e Materials, RC and PC Structures, Masonry Structures. 2013. Italian National Research Council. Available online: https://www.cnr.it/it/node/2620 (accessed on 27 September 2018).

57. *ASCE-SEI 41-06, Seismic Rehabilitation of Existing Buildings, ASCE Standard*; American Society of Civil Engineers: Reston, VA, USA, 2007.

58. Del Zoppo, M.; Di Ludovico, M.; Balsamo, A.; Prota, A.; Manfredi, G. FRP for seismic strengthening of shear controlled RC columns: Experience from earthquakes and experimental analysis. *Compos. Part B* **2017**, *129*, 47–57. [CrossRef]

59. Hak, S.; Morandi, P.; Magenes, G.; Sullivan, T.J. Damage control for clay masonry infills in the design of RC frame structures. *J. Earthq. Eng.* **2012**, *16* (Suppl. 1), 1–35. [CrossRef]

60. Zuccaro, G.; Dolce, M.; De Gregorio, D.; Speranza, E.; Moroni, C. La Scheda CARTIS per la Caratterizzazione Tipologico-Strutturale dei Comparti Urbani Costituiti da Edifici Ordinari. Valutazione Dell'esposizione in Analisi di Rischio Sismico. In Proceedings of the GNGTS 2015. Available online: http://www3.ogs.trieste.it/gngts/files/2015/S23/Riassunti/Zuccaro.pdf (accessed on 27 September 2018). (In Italian)

![buildings logo]

MDPI

Article

Seismic Performance of High-Rise Condominium Building during the 2015 Gorkha Earthquake Sequence

Suraj Malla [1], Sudip Karanjit [2], Purushottam Dangol [1] and Dipendra Gautam [3,*]

[1] Earthquake Resistant Technology Development and Consultancy Center, Kathmandu 44621, Nepal; mallasuraj618@gmail.com (S.M.); strdyn33@gmail.com (P.D.)
[2] Institute of Engineering, Pulchowk Campus, Lalitpur 44700, Nepal; karanjit.sudip@khec.edu.np
[3] Structural and Geotechnical Dynamics Laboratory, StreGa, DiBT, University of Molise, CB 86100 Campobasso, Italy
* Correspondence: dipendra.gautam@unimol.it; Tel.: +39-338-856-9678

Received: 27 November 2018; Accepted: 28 January 2019; Published: 30 January 2019

Abstract: On 25 April 2015, a strong earthquake of magnitude 7.8 struck central Nepal including the capital city, Kathmandu. Several powerful aftershocks of magnitude 6.7, 6.9 and 7.3 together with hundreds of aftershocks of local magnitude greater than 4 hit the same area until May 2015. This earthquake sequence resulted in considerable damage to the reinforced concrete buildings apart from brick and stone masonry constructions. High-rise buildings in Nepal are mainly confined in Kathmandu valley and their performance was found to be in the life safety to collapse prevention level during the Gorkha earthquake sequence. In this paper, seismic performance assessment of a reinforced concrete apartment building with brick infill masonry walls that sustained life safety performance level is presented. Rapid visual assessment performed after the 12 May aftershock (M_W 7.3) highlighted the need for detailed assessment, thus, we carried out nonlinear time history analysis using the recorded accelerograms. The building was first simulated for the recorded acceleration time history (PGA = 0.16 g) and the PGA was scaled up to 0.36 g to assess the behaviour of building in the case of the maximum considered earthquake occurrence. The sum of results and observations highlighted that the building sustained minor damage due to low PGA occurrence during the Gorkha earthquake and considerable damage would have occurred in the case of 0.36 g PGA.

Keywords: seismic performance; high-rise RC; inter-storey drift; Gorkha earthquake

1. Introduction

High-rise apartment building construction in Nepal started mainly after 2000 and most of such constructions are constructed within Kathmandu valley. The density of high-rise construction is greater towards the southern part of Kathmandu valley than in any other parts of the city. As the horizontal expansion of the settlements is almost saturated in Kathmandu valley, medium to high rise buildings are the only option to meet the growing housing demand. Medium to high rise buildings in Nepal are special moment resisting frame (SMRF) constructions designed per Indian Standard Code (ISC) [1]. Before the 2015 Gorkha earthquake, there were 70 high-rise apartments in Kathmandu valley. Among them, two were red tagged (non-habitable) after the Gorkha earthquake and the performance level was 'life safety' in rest of the apartment buildings. Several other studies (e.g., Gautam et al. [2], Gautam and Chaulagain [3], among others) present generic observation reports of various types of existing buildings in Kathmandu valley and other affected areas; however, results of analytical models are limited (e.g., Barbosa et al. [4]). Varum et al. [5] performed ambient vibration measurements in some medium to high-rise buildings in Kathmandu valley after the Gorkha earthquake to have insights of the damages focusing on the effects of infill masonry. Gautam et al. [6] recently formulated the taxonomy and vulnerability of Nepali residential buildings and they placed medium to high rise RC buildings under D-E vulnerability class per EMS-98 vulnerability classification system [7].

Post-earthquake damage assessment and validation efforts are important to understand the behaviour of the structures under extreme loading. Several studies (e.g., [4–9], among others) in the past focused on the same approach used in this study. On the other hand, quite a few studies (e.g., [10–12]) focused on the seismic performance and vulnerability of medium-to-high rise buildings after significant earthquakes in Italy and Spain as well. Similarly, Westenenk et al. [13] conducted field survey and analysed some buildings after the 2010 Maule earthquake in Chile. They concluded that less than 10% elements were damaged due to the Maule earthquake as the structures were not capable of distributing the damage. Scawthon [14] reported the damage in all building typologies after the 2004 Niigata earthquake in Japan and concluded that newer constructions performed better than other building types. Holliday and Grant [15] presented an account of collapsed and survived buildings during the 2010 Haiti earthquake. Similar case studies were also presented by Gautam and Rodrigues [16] for the buildings affected by several earthquakes in Nepal. The Icelandic building damage scenario was presented by Rupakhety et al. [17]. Computational aspects of infill masonry and performance of similar types of buildings are reported by several researchers worldwide (e.g., [18–22]). In most of the contributions, researchers performed damage assessment in the beginning and then analysed the same structures using finite element modelling. Some of the post-earthquake damage assessment studies extended their works to disseminate the vulnerability of building classes as well. Thus, it is clear that post-earthquake damage assessment and subsequent interpretations or modelling are very important to understand the seismic behaviour of typical building classes.

Aiming to depict the seismic performance of a condominium located in Kathmandu, this study presents the details of damage recorded during the field reconnaissance. Thereafter then three-dimensional finite element modelling was done for the same building. Description of the case study block, site condition, strong ground motion and visual damage observation are presented together with the results of nonlinear time history analysis to highlight the seismic performance of condominium buildings in the case of low to moderate PGA occurrence.

2. Materials and Methods

A building from central Kathmandu was selected after the earthquake to study the seismic performance of multi-storey engineered RC building. The building was constructed in 2010. At first, the building was visually inspected using the FEMA-154 guidelines [23] and the rapid visual assessment of the building prompted a detailed vulnerability assessment. Considering the level of damage, a detailed damage assessment was conducted using finite element modelling. Nonlinear pushover and nonlinear time history analysis were adopted in SAP 2000 v.19 [24] to study the behaviour of the case study building. The case study building is a reinforced concrete construction with brick infill masonry walls. The building has a basement, ground floor, six stories and a stair cover (B+G+6+SC). Furthermore, the building comprises 230 mm thick masonry wall throughout. The weight of the infill masonry was also considered during structural modelling. The building has the dimension of 21.5 × 25.5 m in x- and y- directions respectively. Total floor area of the building under study is 5808.37 sq. ft. Similarly, the floor height of the building is 3 m, whereas the basement height is 2.9 m. This condominium building is a representative of the ongoing high-rise construction in Kathmandu valley. Similar construction technology, workmanship and structural as well as non structural components could be found in almost all high-rise apartment buildings as reported by Barbosa et al. [4], thus, this building was considered as a case study building.

The building is situated in a medium soil type as per the Indian Standard Code. The medium soil type was determined using the results of geotechnical logging that was done before the construction of the condominium. The medium soil type indicates poorly graded sands with gravel with little or no fines having standard penetration resistance values between 10 and 30 as per the Indian standard [25].

In the building, M25 grade concrete and Fe-500 reinforcement bars were used during construction. As per the recently updated Indian Standard Code the importance factor of 1.2 was considered for finite element analysis. The response modification factor and damping factor were respectively adopted as 5% and 5% as suggested by the Indian Standard Code [25]. The seismic acceleration coefficient (C_a) was taken as 0.16 and the seismic velocity coefficient (C_v) was taken as 0.2176 for 0.16 g PGA. In this case, the base shear was calculated as 482.051 KN. Furthermore, to assess the performance of the building in the case of design PGA, that is, 0.36 g, another analysis was conducted using C_a as 0.36 and C_v as 0.49 that resulted the total base shear of 675.92 KN. After calculating the base shear, a three-dimensional finite element bare frame model was prepared using SAP 2000 v.19 [24]. To obtain some properties for the validation, non-destructive test was carried out using Rebound Hammer. The test suggested the compressive strength of the concrete as 25 MPa and the same value was adopted in the finite element model as shown in Figure 1. A typical floor plan (ground floor) is shown in Figure 2. Similarly, Figure 3 shows details of a typical beam and a column of the case study building. The stress-strain relationship as suggested by IS 456:2000 [25] was used in modelling. Basic material properties used in modelling are as listed as follows:

Modulus of Elasticity of steel, Es = 200,000 MPa
Modulus of Elasticity of concrete, $E_c = 5000\sqrt{fck} = 25{,}000$ MPa
Characteristic strength of concrete, $f_{ck} = 25$ MPa
Yield stress for steel, $f_y = 500$ MPa

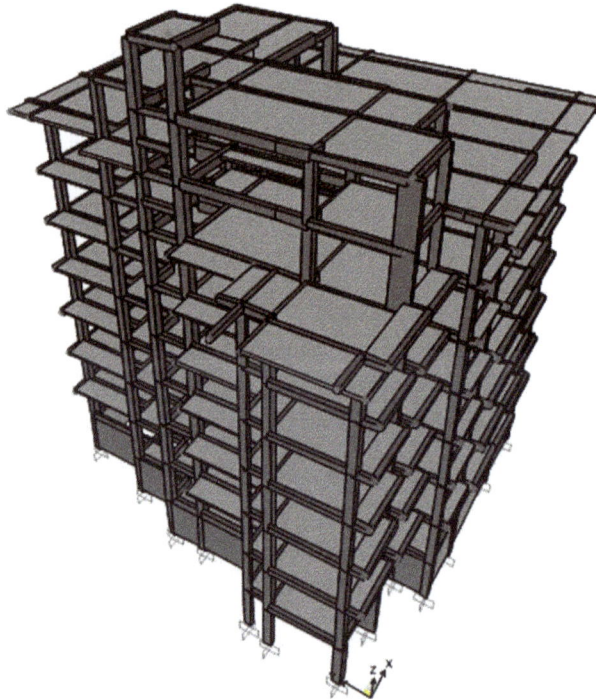

Figure 1. Finite element model of the case study building.

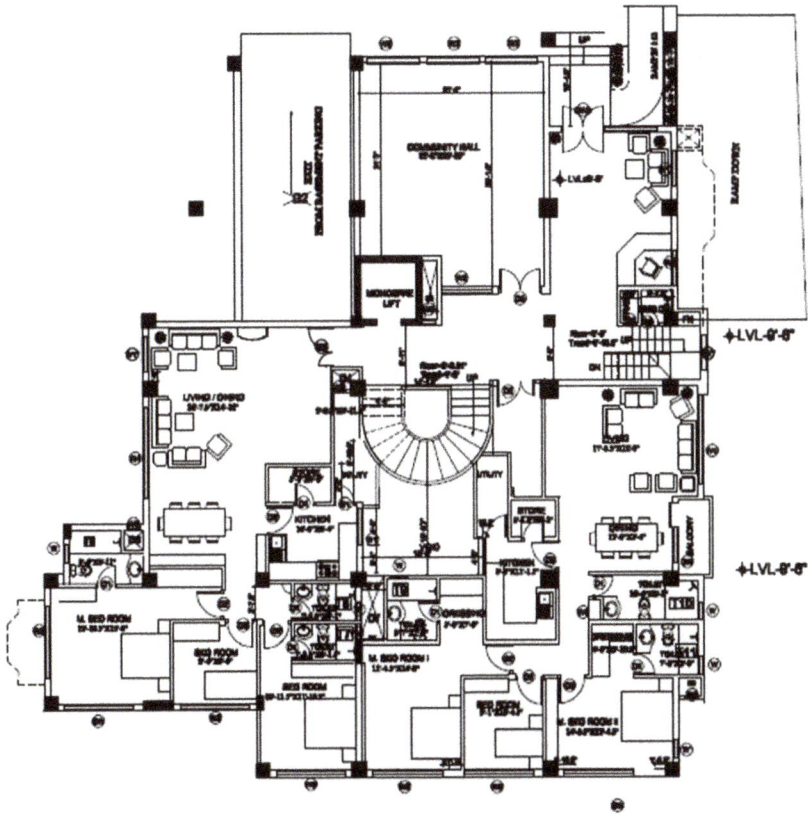

GROUND FLOOR PLAN

Figure 2. Ground floor plan of the case study building.

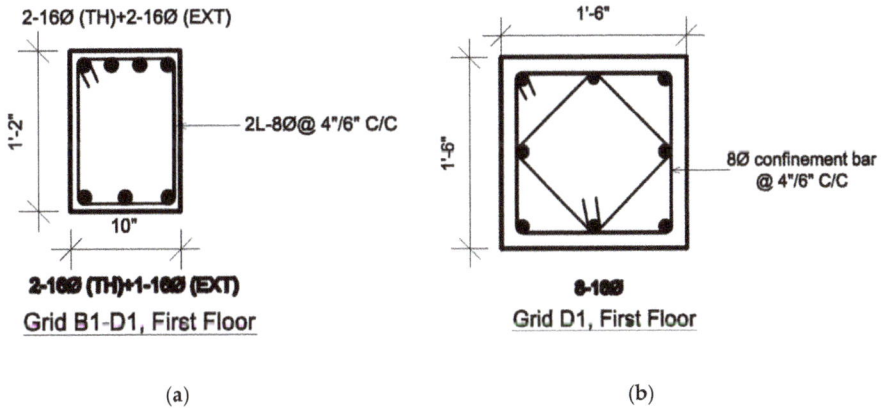

2-16Ø (TH)+2-16Ø (EXT)

1'-6"

2L-8Ø@ 4"/6" C/C

8Ø confinement bar @ 4"/6" C/C

2-16Ø (TH)+1-16Ø (EXT)

8-16Ø

Grid B1-D1, First Floor

Grid D1, First Floor

(**a**)   (**b**)

Figure 3. Details of structural members: (**a**) Typical beam; (**b**) Typical column.

Nonlinear time history analyses were conducted using KATNAP time history recorded by the United States Geological Survey [26]. The acceleration time history used for the time history analysis is shown in Figure 4. Nonlinear analysis was performed using Taketa model which is available in the SAP 2000 finite element program.

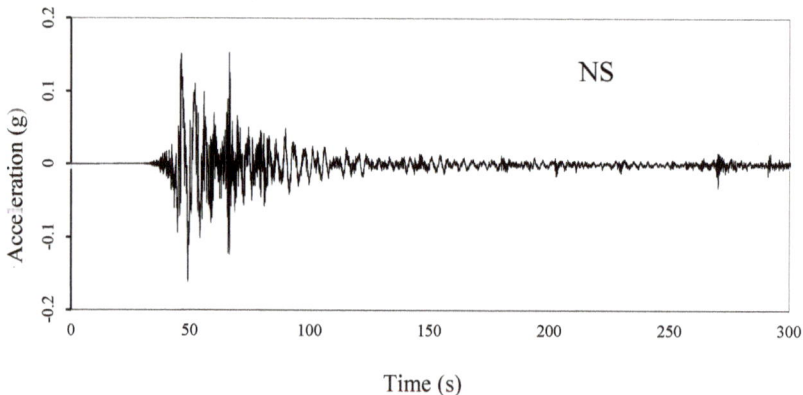

Figure 4. Acceleration time history of Gorkha earthquake (North-South component) at KATNAP station.

3. Rapid Visual Assessment after the Gorkha Earthquake Sequence

Figure 5 shows a field observation sheet recorded during rapid visual observation. As noted in Figure 5, the building is used for residential purpose and the occupancy range was 101–1000. The building is a space frame building containing infill walls. Infill wall was unreinforced brick masonry; thus, the building type was noted as C3 as shown in Figure 5. The building was constructed following the IS code provisions up to the 7th floor with basement and roof cover and from the observations, the case study building was irregular in vertical direction. As the total score (0.1) was less than the cut-off score of 2.0, detailed evaluation of the building was required. To perform the detailed evaluation of the building, detailed survey was conducted for each component of the building. Mostly, infill wall damage was noted throughout the building. Figure 6 shows the infill wall damage which is present throughout the building. Fundamentally, two major reasons would have contributed to the widespread damage to the infill walls. The first is due to lack of reinforcement so that the masonry units were not able to sustain the shaking as in such buildings infills are provided once the frames are finished. Due to prevalent construction practice of providing infill walls once the frame is completed, the structural and non-structural components would not be duly connected. In addition to damage in infill walls; other recorded damages were categorized into seven types: HLC (hairline crack), PC (plaster crack), PF (plaster fall), WC (wall crack/brick crack), WS (wall spill and shift, that is, shifting of wall due to concrete spill), WB (wall and beam joint crack), MJ (mortar joint crack), TC (tile crack and fall). Distribution of these seven types of damages in each storey was recorded carefully and the damage fraction to each category was also calculated. The distribution of various types of damages in each storey is presented in Figure 7. Different types of damages were noted for each storey of the building first and then as per the occurrence of each damage category, the total damage to each storey was converted to 100%. Thereafter, share of each damage mechanism was calculated and finally Figure 7 was plotted. As shown in Figure 7, plaster crack comprised the largest damage fraction throughout the building. In the ground storey, PC, PF and MJ crack were significant damage types observed during the assessment. Similarly, PF, PC, MJ and WC were noted as the significant damage types in the first storey. The second storey had HLC, PC and MJ as the dominant damage types. The third storey observed PC, MJ, TC and WC as the major damage types. The fourth storey depicted the similar damage pattern as that of the third storey. In the fifth storey, PC, WC, WS and MJ were noted as the significant damage types. The sixth storey reflected that the dominant damage types were TC, PC, MJ and WC. It is noted that the wall crack was also dominant in the upper storeys. This should be probably due to the effect of strong vertical shaking during the 2015 Gorkha earthquake as highlighted by Gautam et al. [27] and Gautam and Chaulagain [3].

Rapid Visual Screening of Buildings for Potential Seismic Hazards
FEMA-154 Data Collection Form

HIGH Seismicity

Address: _____
_____ Zip _____
Other Identifiers _____
No. Stories _____ Year Built _____
Screener _____ Date _____
Total Floor Area (sq. ft.) _____
Building Name _____
Use _____

PHOTOGRAPH

Scale:

OCCUPANCY	SOIL	TYPE	FALLING HAZARDS
Assembly Govt Office	Number of Persons	A B C D E F	☐ ☐ ☐ ☐
Commercial Historic (Residential)	0 – 10 11 – 100	Hard Avg. Dense Stiff Soft Poor	Unreinforced Parapets Cladding Other:
Emer. Services Industrial School	(101-1000) 1000+	Rock Rock Soil Soil Soil Soil	Chimneys

BASIC SCORE, MODIFIERS, AND FINAL SCORE, S

BUILDING TYPE	W1	W2	S1 (MRF)	S2 (BR)	S3 (LM)	S4 (RC SW)	S5 (URM INF)	C1 (MRF)	C2 (SW)	C3 (URM INF)	PC1 (TU)	PC2	RM1 (FD)	RM2 (RD)	URM
Basic Score	4.4	3.8	2.8	3.0	3.2	2.8	2.0	2.5	2.8	1.6	2.6	2.4	2.8	2.8	1.8
Mid Rise (4 to 7 stories)	N/A	N/A	+0.2	+0.4	N/A	+0.4	+0.4	+0.4	+0.4	+0.2	N/A	+0.2	+0.4	+0.4	0.0
High Rise (> 7 stories)	N/A	N/A	+0.6	+0.8	N/A	+0.8	+0.8	+0.6	+0.8	+0.3	N/A	+0.4	N/A	+0.6	N/A
Vertical Irregularity	-2.5	-2.0	-1.0	-1.5	N/A	-1.0	-1.0	-1.5	-1.0	-1.0	N/A	-1.0	-1.0	-1.0	-1.0
Plan Irregularity	-0.5	-0.5	-0.5	-0.5	-0.5	-0.5	-0.5	-0.5	-0.5	-0.5	-0.5	-0.5	-0.5	-0.5	-0.5
Pre-Code	0.0	-1.0	-1.0	-0.8	-0.6	-0.8	-0.2	-1.2	-1.0	-0.2	-0.8	-0.8	-1.0	-0.8	-0.2
Post-Benchmark	+2.4	+2.4	+1.4	+1.4	N/A	+1.6	N/A	+1.4	+2.4	N/A	+2.4	N/A	+2.8	+2.6	N/A
Soil Type C	0.0	-0.4	-0.4	-0.4	-0.4	-0.4	-0.4	-0.4	-0.4	-0.4	-0.4	-0.4	-0.4	-0.4	-0.4
Soil Type D	0.0	-0.8	-0.6	-0.6	-0.6	-0.6	-0.4	-0.6	-0.6	-0.4	-0.6	-0.6	-0.6	-0.6	-0.6
Soil Type E	0.0	-0.8	-1.2	-1.2	-1.0	-1.2	-0.8	-1.2	-0.8	-0.8	-0.4	-1.2	-0.4	-0.6	-0.8

FINAL SCORE, S	0.1

COMMENTS	Detailed Evaluation Required
	(YES) NO

* = Estimated, subjective, or unreliable data
DNK = Do Not Know

BR = Braced frame
FD = Flexible diaphragm
LM = Light metal

MRF = Moment-resisting frame
RC = Reinforced concrete
RD = Rigid diaphragm

SW = Shear wall
TU = Tilt up
URM INF = Unreinforced masonry infil

Figure 5. Summary of the rapid visual assessment of the building.

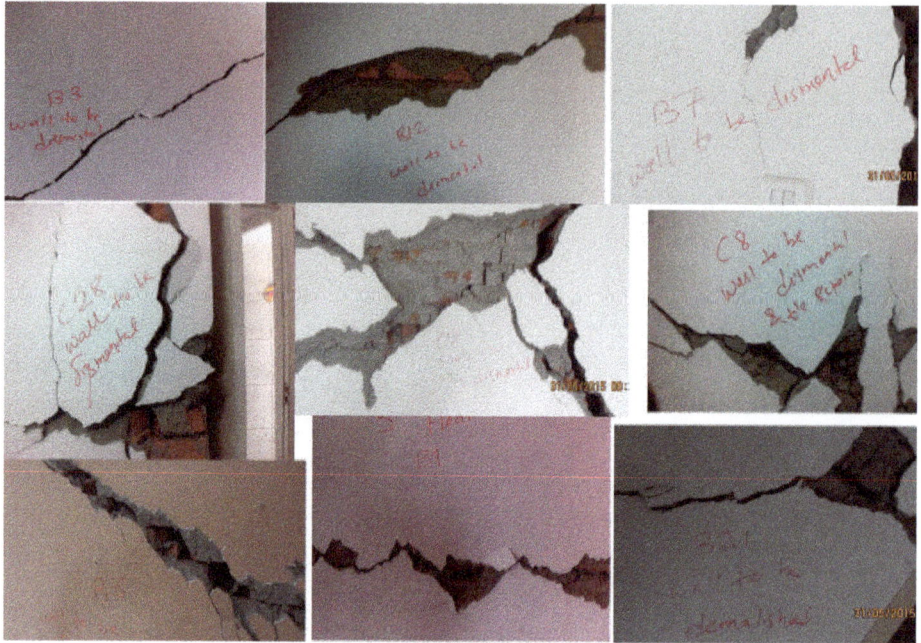

Figure 6. Infill wall damage due to Gorkha earthquake in the case study building.

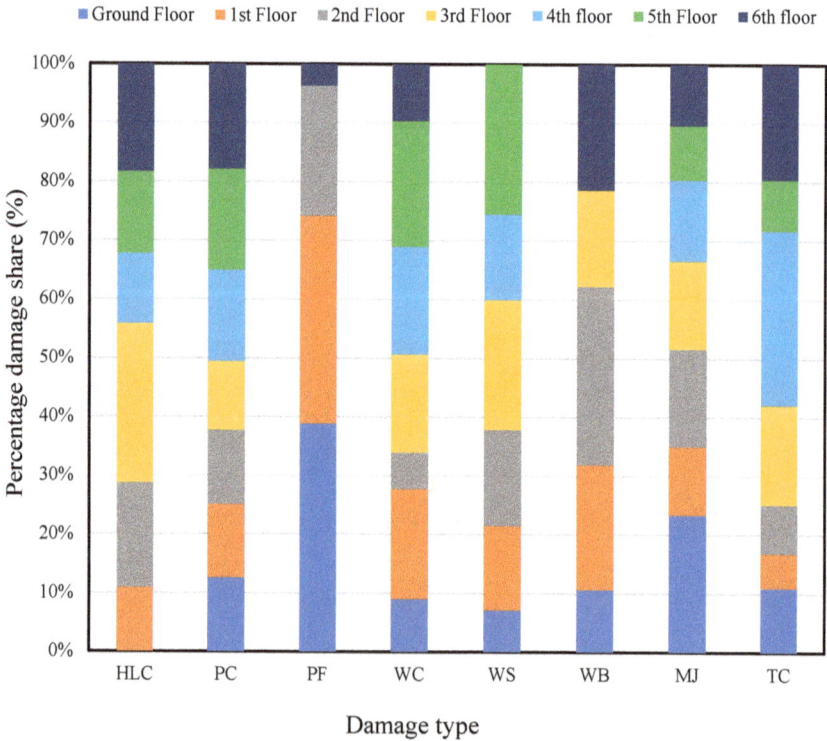

Figure 7. Storey-wise distribution of damage in the case study building.

4. Analytical Modelling Results

Pushover analyses in the x- and y- directions were performed at first and the pushover curves of the building are shown in Figure 8. As shown in Figure 8, base shear capacity of the case study building can be considered significantly high. The seismic weight of the building was calculated as 51,177.75 KN considering 100% of dead load (DL), 25% of live load (LL) (when LL $\leq$ 3 KN/m^2) and 50% of live load (LL) (when LL > 3 KN/m^2). The natural periods of the building in x- and y-directions were obtained as 0.588 s and 0.551 s respectively. To perform pushover analysis, 11 diaphragms were created for the entire structure as shown in Table 1. After assigning the hinges, the storey drifts were obtained from the analysis for both 0.16 g (PGA equivalent to Gorkha earthquake) and 0.36 g (PGA equivalent to the maximum considered earthquake, that is, MCE) PGAs. Both pushover analysis and nonlinear time history analyses were conducted to check the variation of performance of the structure. Figure 9 depicts the inter-storey drift (ISD) plots for pushover and time history analysis for the corner column as well as the centre of mass (CM). As shown in Figure 9, the maximum drift was observed in the second floor (diaphragm 3–4) in the case of time history analysis along the x-axis. Figure 9 also highlights that the pushover analyses leads to conservative results than the time history in terms of inter-storey drift. As the maximum inter-storey drift was limited to 1.2% that falls under 'operational' performance level per VISION-2000 [28]. The same performance level was observed during the Gorkha earthquake as well (see e.g., [4]). One more analysis was conducted to depict the behaviour of the building in the case of MCE. Thus, the PGA was scaled up to 0.36 g and both pushover and nonlinear time history analyses were carried out in both x- and y-directions. Using the similar hinge conditions, the ISD for 0.36 g PGA is presented in Figure 10.

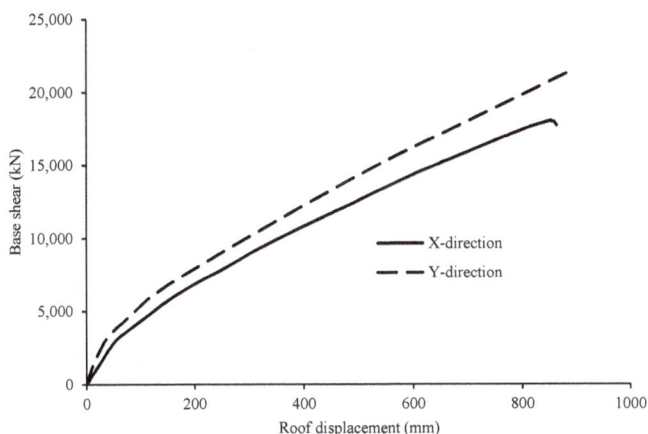

Figure 8. Pushover curves for the building.

Table 1. Configuration of diaphragm for FE analysis.

Diaphragm	Floor level
9-10	Stair Cover
8-9	7th Floor
7-8	6th Floor
6-7	5th Floor
5-6	4th Floor
4-5	3rd Floor
3-4	2nd Floor
2-3	1st Floor
1-2	Ground Floor
0-1	Basement
0	Base

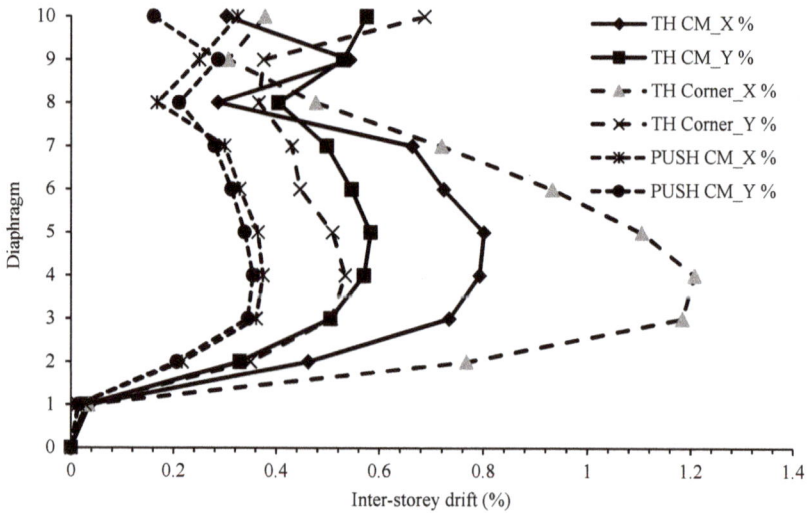

Figure 9. Inter-storey drift ratio for the 0.16 g PGA input. In figure, TH indicates the time history analysis and PUSH indicates the pushover analyses conducted in both x- and y-directions.

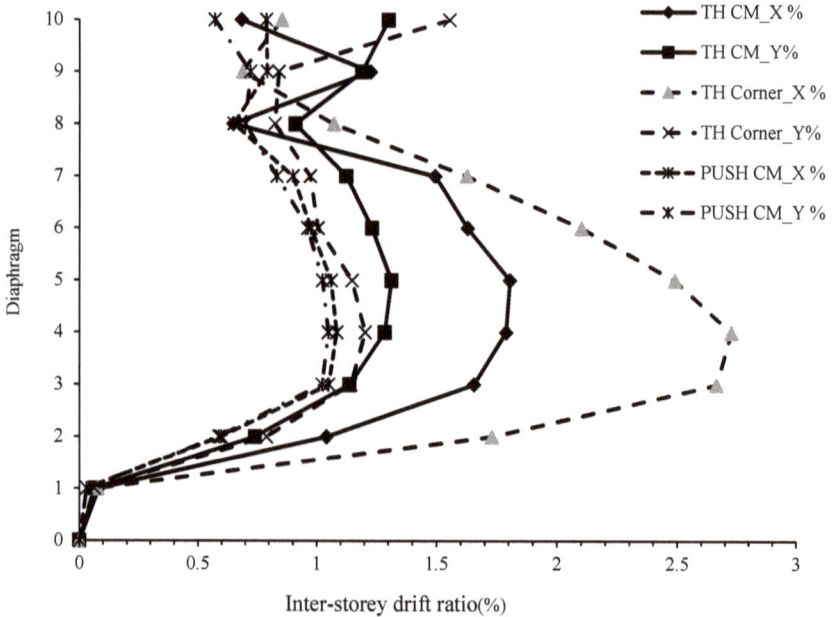

Figure 10. Inter-storey drift ratio for the 0.36 g PGA input. In figure, TH indicates the time history analysis and PUSH indicates the pushover analyses conducted in both x- and y-directions.

As shown in Figure 10, the maximum ISD has reached greater than 2.5% which denotes that in the case of MCE, the building would cross the life safety performance level. So, the moderate damage recorded during the Gorkha earthquake should be clearly due to the low PGA occurrence in Kathmandu valley. As in the case of 0.16 g PGA, pushover analyses depicted conservative results in terms of the inter-storey drift at 0.36 g too.

5. Conclusions

After the 2015 Gorkha earthquake sequence, rapid visual assessment was conducted in an apartment building located in central Kathmandu, Nepal. The rapid visual screening suggested the detailed assessment, thus, detailed damage assessments were conducted in two phases. The first tranche of assessment comprised damage identification and the second tranche comprised finite element modelling of the structure that covered pushover and nonlinear time history analysis. Modelling was done at 0.16 g PGA (as recorded during the Gorkha earthquake) and 0.36 g (which is equivalent to the maximum considered earthquake PGA). As the PGA was relatively low during the Gorkha earthquake, the building showed operational performance level; however, at 0.36 g, the results indicate that the storey drift would cross the life safety performance level leading to a severe damage to the building. It is fundamental to note that the damage was concentrated in non-structural component of the building. More precisely, the damage was primarily in the brick infills throughout the building.

One of the limitations of this study is that infill models were not created but the weight of the wall was considered in analysis so that the exact mechanisms in each infill wall were not cross-validated. It is worthy to note that the contribution of infill panels is a crucial aspect in the seismic performance of building. Owing to this fact and limitation of this study, readers are directed to the related papers to have further insights (e.g., [18–22]). Future studies may incorporate the contribution of infill walls to justify the damage occurrence and seismic performance of the building.

Author Contributions: Conceptualization, S.M.; methodology, D.G. and S.K.; software, S.M., S.K. and D.G.; validation, S.K. and D.G.; formal analysis, S.M.; investigation, S.M.; resources, P.D., D.G. and S.K.; data curation, S.M.; writing—original draft preparation, D.G.; writing—review and editing, D.G., S.K., P.D. and S.M.; visualization, D.G. and S.M.; supervision, S.K. and P.D.; project administration, P.D.

Funding: This research received no external funding.

Acknowledgments: Authors are grateful to the authorities who allowed to investigate the building.

Conflicts of Interest: The authors declare no conflict of interest.

References

1. Bureau of Indian Standard. *Criteria for Earthquake Resistant Design of Structures*, 6th ed.; Bureau of Indian Standard: New Delhi, India, 2016.
2. Gautam, D.; Rodrigues, H.; Bhetwal, K.K.; Neupane, P.; Sanada, Y. Common structural and construction deficiencies of Nepalese buildings. *Innov. Infrastruct. Solut.* **2016**, *1*. [CrossRef]
3. Gautam, D.; Chaulagain, H. Structural performance and associated lessons to be learned from world earthquakes in Nepal after 25 April 2015 (M_W 7.8) Gorkha earthquake. *Eng. Fail. Anal.* **2016**, *68*, 222–243. [CrossRef]
4. Barbosa, A.R.; Fahnestock, L.A.; Fick, D.R.; Gautam, D.; Soti, R.; Wood, R.; Moaveni, B.; Stavridis, A.; Olsen, M.J.; Rodrigues, H. Performance of medium-to-high rise reinforced concrete frame buildings with masonry infill in the 2015 Gorkha, Nepal, earthquake. *Earthq. Spectra* **2017**, *33*, S197–S218. [CrossRef]
5. Varum, H.; Furtado, A.; Rodrigues, H.; Dias-Oliveira, J.; Vila-Pouca, N.; Arede, A. Seismic performance of the infill masonry walls and ambient vibration tests after the Ghorka 2015, Nepal earthquake. *Bull. Earthq. Eng.* **2017**, *15*, 1185–1212. [CrossRef]
6. Gautam, D.; Fabbrocino, G.; Santucci de Magistris, F. Derive empirical fragility functions for the residential buildings in Nepal. *Eng. Struct.* **2018**, *171*, 612–628. [CrossRef]
7. Grunthal, G. *Euripean Macroseismic Scale 1998 (EMS-98)*; Centre Européen de Géodynamique et de Séismologie: Luxemburg, 1998.
8. Centeno, J.; Ventura, C.E.; Ingham, J.M. Seismic performance of a six-story reinforced concrete masonry building during the Canterbury earthquake sequence. *Earthq. Spectra* **2014**, *30*, 363–381. [CrossRef]

9. Sawaki, Y.; Rupakhety, R.; Olafsson, S.; Gautam, D. System identification of a residential building in Kathmandu using aftershocks of 2015 Gorkha earthquake and triggered noise data. In Proceedings of the International Conference on Earthquake Engineering and Structural Dynamics, ICESD 2017; Rupakhety, R., Olafsson, S., Bessason, B., Eds.; Geotechnical, Geological and Earthquake Engineering. Springer: Berlin, Germany, 2019; Volume 47, pp. 233–247. [CrossRef]
10. Verderame, G.M.; Ricci, P.; De Luca, F.; Del Gaudio, C.; De Risi, M.T. Damage scenarios for RC buildings during the 2012 Emilia (Italy) earthquake. *Soil Dyn. Earthq. Eng.* **2014**, *66*, 385–400. [CrossRef]
11. Del Gaudio, C.; Ricci, P.; Verderame, G.M.; Manfredi, G. Urban scale seismic fragility assessment of RC buildings subjected to L'Aquila earthquake. *Soil Dyn. Earthq. Eng.* **2017**, *96*, 47–63. [CrossRef]
12. Romao, X.; Costa, A.A.; Pauperio, E.; Rodrigues, H.; Vicente, R.; Varum, H.; Costa, A. Field observations and interpretation of the structural performance of constructions after the 11 May 2011 Lorca earthquake. *Eng. Fail. Anal.* **2013**, *34*, 670–692. [CrossRef]
13. Westenenk, B.; de la Llera, J.C.; Besa, J.J.; Junemann, R.; Moehle, J.; Luders, C.; Inaudi, J.A.; Elwood, K.J.; Hwang, S.-H. Response of reinforced concrete buildings in Concepcion during the Maule earthquake. *Earthq. Spectra* **2012**, *28*, S257–S280. [CrossRef]
14. Scawthorn, C. Building aspects of the 2004 Niigata Ken Chuetsu, Japan, earthquake. *Earthq. Spectra* **2006**, *22*, S75–S88. [CrossRef]
15. Holliday, L.; Grant, H. Haiti building failures and a replicable building design for improved earthquake safety. *Earthq. Spectra* **2011**, *27*, S277–S297. [CrossRef]
16. Gautam, D.; Rodrigues, H. Seismic vulnerability of urban vernacular buildings in Nepal: Case of Newari construction. *J. Earthq. Eng.* **2018**. [CrossRef]
17. Rupakhety, R.; Sigbjornsson, R.; Olafsson, S. Damage to residential buildings in Hveragerði during the 2008 Ölfus Earthquake: Simulated and surveyed results. *Bull. Earthq. Eng.* **2016**, *14*, 1945–1955. [CrossRef]
18. Gerardi, V.; Ditommaso, R.; Auletta, G.; Ponzo, F.C. Reinforced concrete framed structures: Numerical validation of two physical models capable to consider the stiffness contribution of infill panels on framed structures in operative conditions. *Ing. Sismica* **2018**, *35*, 1–21.
19. Bilotta, A.; Tomeo, R.; Nigro, E.; Manfredi, G. Evaluation of the seismic demand of an existing tall building. *Ing. Sismica* **2018**, *35*, 67–87.
20. Montuori, R.; Nastri, E.; Piluso, V. Modelling of floor joists contribution to the lateral stiffness of RC buildings designed for gravity loads. *Eng. Struct.* **2016**, *121*, 85–96. [CrossRef]
21. Donà, M.; Minotto, M.; Saler, E.; Tecchio, G.; da Porto, F. Combined in-plane and out-of-plane seismic effects on masonry infills in RC frames. *Ing. Sismica* **2017**, *34*, 157–173.
22. Preti, M.; Bolis, V. Seismic analysis of a multistory RC frame with infills partitioned by sliding joints. *Ing. Sismica* **2017**, *34*, 175–187.
23. Federal Emergency Management Agency. *Rapid Visual Screening of Building for Potential Hazards: A Handbook*; Federal Emergency Management Agency: Washington, DC, USA, 2015.
24. Computers and Structures [CSI] Inc. *SAP: Integrated Software for Structural Analysis and Design, v. 19*; Computers and Structures [CSI] Inc.: Walnut Cree, CA, USA, 2001.
25. Bureau of Indian Standard. *Plain and Reinforced Concrete–Code of Practice [IS 456-2000]*; Bureau of Indian Standard: New Delhi, India, 2000.
26. U.S.G.S. M 7.8–36 km E of Khudi, Nepal. 2015. Available online: https://earthquake.usgs.gov/earthquakes/eventpage/us20002926/executive (accessed on 25 November 2018).
27. Gautam, D.; Santucci de Magistris, F.; Fabbrocino, G. Soil liquefaction in Kathmandu valley due to 25 April 2015 Gorkha, Nepal earthquake. *Soil Dyn. Earthq. Eng.* **2017**, *97*, 37–47. [CrossRef]
28. Structural Engineers Association of California (SEAOC), VISION-2000. *A Framework for Performance-Based Design*; Vols. I, II, III; VISION-2000 Committee: Sacramento, CA, USA, 1995.

![buildings logo] **buildings**

MDPI

Article

Discrete Element Modeling of the Seismic Behavior of Masonry Construction

José V. Lemos

Laboratório Nacional de Engenharia Civil (LNEC), Av. do Brasil 101, 1700-066 Lisboa, Portugal; vlemos@lnec.pt; Tel.: +351-218443000

Received: 31 December 2018; Accepted: 5 February 2019; Published: 10 February 2019

Abstract: Discrete element models are a powerful tool for the analysis of masonry, given their ability to represent the discontinuous nature of these structures, and to simulate the most common deformation and failure modes. In particular, discrete elements allow the assessment of the seismic behavior of masonry construction, using either pushover analysis or time domain dynamic analysis. The fundamental concepts of discrete elements are concisely presented, stressing the issues related to masonry modeling. Methods for generation of block models are discussed, with some examples for the case of irregular stone masonry walls. A discrete element analysis of a shaking table test performed on a traditional stone masonry house is discussed, as a demonstration of the capabilities of these models. Practical application issues are examined, namely the computational requirements for dynamic analysis.

Keywords: masonry structures; numerical modeling; discrete elements; seismic analysis

---

1. Introduction

The safety assessment of masonry structures under seismic loads demands numerical models capable of representing appropriately the response and the failure modes observed during earthquakes. Laboratory testing, namely using shaking tables, offers a controlled environment in which the main variables can be more accurately measured, providing important data for the validation of numerical models. A wide range of models is available nowadays, each one with its specific strengths and a preferred range of applications [1]. Discrete element (DE) models, involving the representation of masonry by a system of interacting distinct blocks, are one of the numerical tools particularly suited for the simulation of phenomena such as sliding and separation along joints, which may induce progressive structural damage and collapse [2].

Many applications of DE models to masonry structures under seismic loading have been reported in the literature. Early works addressed mainly the study of historical monuments involving systems with a relatively small number of blocks [3]. Presently, it is possible to address much more complex structures, composed of a large number of blocks, and assess their seismic capacity by means of either pushover methods [4] or dynamic analysis [5]. DE models using these alternative analysis techniques, based on static or dynamic representations of the seismic action, have been applied in the simulation of lab experiments of masonry walls under out-of-plane loading [6,7]. The failure modes observed in lab tests involving in-plane cyclic loading of masonry panels have also been effectively reproduced by these models [8].

In the literature, several discontinuum modeling techniques have been developed for masonry analysis, which share many of the fundamental concepts of DE models. For example, the rigid body and spring model [9] involves the representation of masonry by a system of rigid blocks, which do not necessarily correspond to the masonry units, as a homogenization procedure is invoked to obtain the contact properties. The applied element method [10,11] also concentrates the deformation at the interfaces between rigid blocks, which are discretized into a fine mesh of contact springs with nonlinear behavior. The fiber contact element method [12] is a related approach, involving the definition of contact properties from a system of fibers that connect adjacent blocks. Within a finite element framework, models based on rigid elements and nonlinear joints have also been advanced for masonry structures, namely resorting to limit analysis concepts [13,14].

In the present article, the fundamental concepts of discrete element models are concisely discussed in the next section, including computational efficiency issues. The generation of models to represent traditional stone wall structures is then addressed. Two different methods of creating irregular assemblies are described, one based on a Voronoi tessellation and the other on overlaying irregular courses. The responses of models generated with different random geometric parameters are compared. The issue of validation of the numerical models is examined, with particular reference to the results of a shaking table test of a physical model of a traditional stone masonry house, for which a simplified block representation was created. Finally, some concluding remarks are presented.

2. Essential Concepts in Discrete Element Modeling

2.1. Block Representation

Discrete element models provide a discontinuous idealization of masonry (commonly designated as micro-modeling), in which the units and the joints are explicitly represented [1]. There are multiple DE formulations and codes available nowadays, using different terminologies, but they all share common underlying concepts [2]. The units are typically represented by polygonal or polyhedral shapes. The simplest models assume the blocks to be rigid bodies, with all the system deformation placed in the joints. This is a good assumption for hard stone blocks, but it also simplifies the analysis of large structures. Deformable blocks provide a more versatile framework. The block deformation is typically modelled by discretizing it internally into a finite element mesh. The element material is assumed elastic in most cases, but nonlinear constitutive models may also be assigned to the elements. For many problems in which the inter-block movements are dominant, the use of rigid blocks provides results practically equivalent to coarse-mesh deformable blocks. A comparative study for out-of-plane failure of walls is presented in [4].

2.2. Contact between Blocks

Most discrete element models adopt a simplified representation of the contact between blocks based on sets of point contacts. Therefore, no joint or interface elements are defined, contrary to finite element micro-models. The point contact approach is more versatile and facilitates the analysis in the large displacement range, as the system connectivity changes, and the type, location, and orientation of contacts need to be periodically updated. The drawback is that more contact points are required to achieve an accurate contact stress distribution.

The point contact approach assumes that the stress at a contact point is a function of the relative displacement of the two blocks at that point. The contact points are typically placed at vertex–face or edge–edge interactions. For deformable blocks, additional contact points are created for the nodal points of the internal element mesh that fall on block boundaries. Block faces are typically discretized into triangular facets, allowing contact areas to be assigned to each contact point, and therefore to relate contact stresses and forces.

An example of contact representation in the code 3DEC [15] is shown in Figure 1. Two types of elementary or point contacts may exist: vertex-to-face (VF) and edge-to-edge (EE). A contact point of VF type is created where a vertex touches a face, while a contact point of EE type is located where two edges intersect. Vertex-to-edge or vertex-to-vertex interactions are considered particular cases of the VF type. The common case of face-to-face interaction between two blocks is represented by several elementary contacts (Figure 1a). In this case, the sum of the areas of the point contacts equals the total area of contact between the blocks. On the other hand, when the blocks interact by two intersecting edges, a case of a true point contact, then a single elementary contact exists. For practical purposes, a true point interaction obviously has a finite stiffness, so a minimum contact area has to be defined, typically a small fraction of the average contact area.

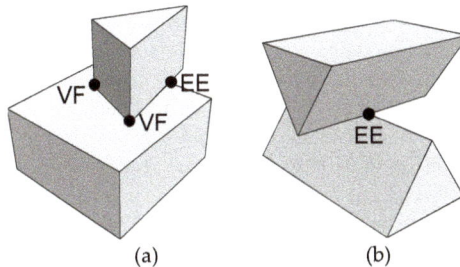

Figure 1. Representation of block interactions by elementary point contacts of vertex-to-face type (VF) and edge-to-edge type (EE): (**a**) Face-to-face block contact; (**b**) Edge-to-edge block contact.

2.3. Solution Methods and Computational Efficiency

The need to represent large displacement problems, demanding a progressive update of block positions and contact point locations, has prompted many DE codes to avoid matrix solutions, opting for explicit solution methods. For time domain analysis, the integration of the equations of motion, of either the rigid blocks or the deformable block nodes, using a central-difference algorithm, is a common choice. A feature of many DE codes is to use the same algorithm for static problems by means of dynamic relaxation, which resorts to artificial damping to obtain convergence to the equilibrium solution, or to a failure mechanism. Adaptive damping and mass scaling techniques may be employed to improve the convergence rate of the relaxation procedure.

These explicit time stepping algorithms are quite effective for quasi-static analysis, namely in the case of pushover methods for seismic assessment. For dynamic analysis, however, since mass scaling cannot be applied, time steps are often small due to numerical stability requirements [9]. These limitations may be particularly severe when the stiffness-proportional component of Rayleigh damping is used. Therefore, run times can be demanding for large or complex models. The use of rigid blocks often allows larger time steps, so they are frequently applied in dynamic analysis, as done in the study described in Section 4.

2.4. Material Properties

The calibration of numerical models with experimental data is a key issue addressed by many authors. For example, a consistent method has been developed to identify the parameters of a DE model of a brick panel from laboratory tests [16]. More often, the calibration is based on empirical trial-and-error procedures until an acceptable fit of key indicators is reached. Parametric studies are invaluable to find the potential influence of material properties that display a significant uncertainty.

A model for dynamic analysis of existing structures under earthquake loading has to provide realistic natural frequencies. Given the uncertainty in assigning the deformability to many types of masonry, particularly for historical structures, in situ measurements by ambient vibration are important to calibrate the model. For a rigid block representation, the model deformability is given by the joint stiffness parameters, which can be estimated directly given the measured frequencies. For example, even for simple stone masonry structures, the estimation of the displacements caused by an earthquake has been shown to be much improved if the structural frequencies are correctly matched [17].

3. Representation of the Block Structure

3.1. Modeling the Masonry Morphology

Devising a block system to represent modern brick masonry poses no difficulty, as only regular shaped blocks are needed, and the procedure can be easily automated. In the case of existing stone masonry structures, the model generation tasks are much more time consuming. We should keep in mind that a numerical model is always an idealization of the real structure. In most cases, the analyst selects a simplified structure in such a way that the key features of the block structure are reproduced. Numerical experiments can be performed to evaluate the effect that a given simplification has on the results, and if this is acceptable given the purpose of the analysis.

In the case of monumental structures composed of large stone blocks, a photo survey may allow the definition of a numerical model that closely approximates the real morphology [18]. However, in most DE models of stone masonry walls, an equivalent regular block system is employed, which respects the typical values of block sizes and the amount of interlocking. In the next two sections, the generation of block systems with irregular geometries is addressed for the case of masonry walls. In order to simplify the issue, only the non-regular pattern visible in the wall plane is modelled. In reality, wall cross-sections are often composed of two or three leaves of block and fill material. A 3D representation of such structures is only feasible in small models of masonry components. When 2D models are employed to study the out-of-plane stability of such walls, however, it is possible to achieve a much more detailed representation. De Felice [19] studied a three-leaf cross section of a traditional wall, simulating closely the observed irregular block geometries, including the inner leaf of rubble masonry.

3.2. Regular and Voronoi Block Patterns

In this study, the influence of the block pattern on the out-of-plane stability of stone masonry walls was analyzed with DE models. The regular block pattern is shown in Figure 2a [20]. The wall dimensions were 20×10 m, with 0.80 m thickness. A Young's modulus of 2.5 GPa was assumed. As the blocks were assumed rigid, the joint stiffnesses were calculated to reproduce the global deformability: normal stiffness of 2.5 GPa/m, and shear stiffness of 1.0 GPa/m. For simplicity, a Coulomb friction model was adopted for the joints, with a friction angle of $35°$, and no cohesion or tensile strength.

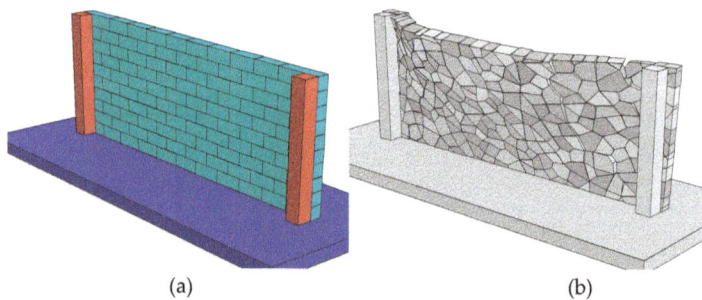

(a)   (b)

Figure 2. Masonry wall models. (**a**) Regular block pattern; (**b**) Voronoi block pattern, showing failure mode for out-of-plane loading.

The rigid base block and the two vertical columns were fixed in all directions (Figure 2a). A distributed static load was applied to the wall in the out of plane direction, progressively increasing until failure ensued. Therefore, the wall was simply supported by the vertical column blocks, approximately simulating the effect of cross-walls, thus allowing the wall ends to rotate.

In addition to the regular system of Figure 2a, Voronoi polygon patterns were randomly created, with the same average block size as the regular mesh. One of these is shown in Figure 2b, at the stage in which the out-of-plane failure mode developed.

Figure 3 compares the force-deformation curves and failure loads of the simulations with 3 different Voronoi block systems, with the same average size. It is interesting to observe that the three randomly generated patterns displayed fairly similar behaviors. In the same graphic, 2 cases of regular joints are shown. To be comparable with the Voronoi model, square blocks were used. The case of no offset had continuous vertical joints. The loss of interlocking led to a lower strength. An offset of 0.1 m clearly increased the initial stiffness and failure load, which was only marginally lower than the 0.5 m offset case shown in Figure 2a. Cross-joint imbrication was critical, and the Voronoi pattern did not create discontinuous joints, so it tended to underestimate the block interlocking.

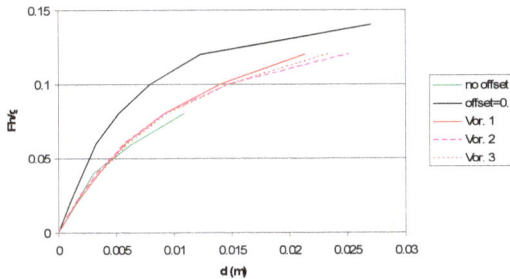

Figure 3. Failure load for wall models: regular model of square blocks with cross-joint offsets of 0 and 0.1 m, and three random Voronoi patterns.

3.3. Irregular Coursed Block Patterns

The Voronoi pattern does not represent the typical coursed structure of traditional stone masonry, being essentially an expeditious way to create randomly shaped block systems. An alternative procedure was proposed to generate block patterns which displayed overlaid courses, but with irregular horizontal joints, course height, and cross-joint imbrication [20], as shown in Figure 4.

Figure 4. Masonry wall model with irregular coursed structure.

The procedure developed to obtain this block pattern involved the sequential generation of the masonry courses using the five geometric parameters shown in Figure 5 [20]. Each one of these parameters was defined in statistical terms by a mean value (m) and a deviation (d). The geometry of the bed joints was composed of a series of segments according to the following parameters: spacing (s_m, s_d); segment length (t_m, t_d); and vertical deviation (h_m, h_d) from the mean trace. In the case shown, for simplicity, a uniform distribution was assumed. For instance, the spacing of bed joints was given by a random number in the interval [$s_m - s_d$, $s_m + s_d$]. The cross joint were introduced sequentially, with each location defined by a spacing parameter (b_m, b_d), and an angle deviation from the vertical

(a_m, a_d). The system in Figure 4 was created with the values: $s_m = 1$ m, $s_d = 0.1$ m; $t_m = 2$ m; $t_d = 1$ m; $h_m = 0$; $h_d = 0.2$ m; $b_m = 1.5$ m; $b_d = 0.5$ m; $a_m = 0$; and $a_d = 0$.

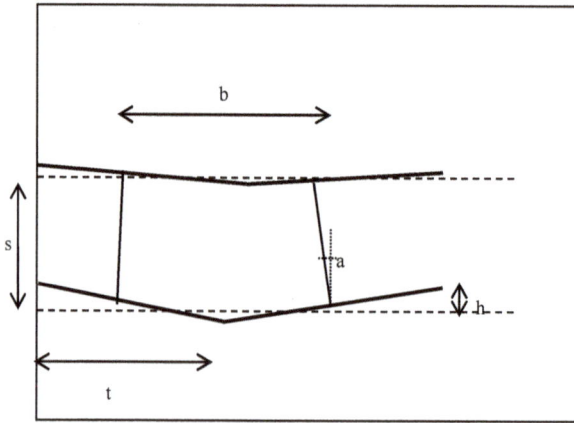

Figure 5. Definition of geometric parameters in block generation procedure.

The system in Figure 4 was one of three randomly created, and their force–displacement curves are compared in Figure 6 with those from regular jointed models with rectangular blocks (Figure 2a). The 3 irregular patterns displayed failure loads in the same range, which were also close to the case of the regular pattern with the smallest offset (0.1 m), and well above the curve for the continuous cross joints. It should be noted that the regular block failure loads in this figure, with rectangular shapes of ratio 2:1, were higher than the ones in Figure 3, which corresponded to square blocks.

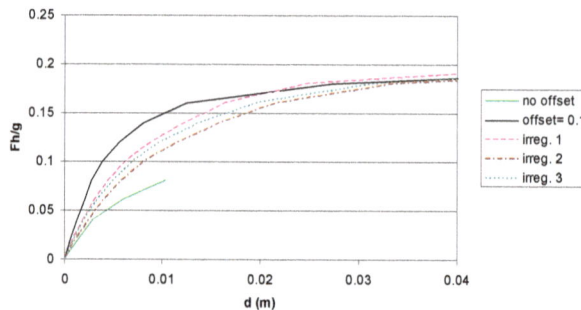

Figure 6. Failure load for wall models: regular model of rectangular blocks with cross-joint offsets of 0 and 0.1 m, and three irregular coursed patterns.

4. DEM Analysis of a Shaking Table Test of a Stone Masonry Model

4.1. Experimental study

Shake table testing provides comprehensive data on the behavior of masonry structures under intense dynamic loading, which is invaluable for the validation of numerical codes. In the framework of the 9th International Masonry Conference, which took place in Guimarães, a comparative study of different numerical models was undertaken, involving blind predictions and a posteriori analyses of shake table tests on brick and stone masonry models performed at LNEC [21,22]. The stone masonry structure, shown in Figure 7, consisted of a main gabled wall, with a door opening, and return walls on both ends. An opening was placed in one of the return walls, resulting in an asymmetry, inducing torsional movements. The walls, built of irregular stones, each had a thickness of 0.5 m. Unidirectional ground motion was applied in the out-of-plane direction of the main wall. Various finite element and discrete element models were employed in the comparative study [22]. The main issues involved in the DE representation of the experiment, and the lesson learned, are discussed in the following section, with particular reference to the 3DEC rigid block analyses [23]. Two other types of DE models were also included in the study: a combined finite–discrete element approach [24]; and a macro-element formulation [25].

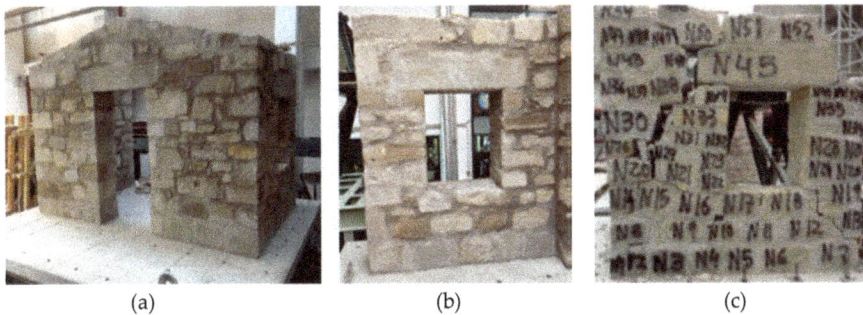

Figure 7. Stone structure: (**a**) general view and (**b**) return wall with opening; (**c**) damage of the return wall after the test (adapted from [23]).

4.2. DE Representation of the Stone Masonry Test Model

DE allows various levels of detail to be included in the representation of a masonry structure. It would be possible to devise a model in which each real stone would be simulated, adopting a more or less complex geometry. However, in the analyses of the stone masonry specimen performed with 3DEC, a simplified representation was chosen [23]. The aim was to follow a model generation methodology applicable to the practical study of a real structure, where the real block shapes are known. Only the main pattern of the stone arrangement was attempted, respecting the average unit dimensions, but assuming horizontal joints, and vertical joint offsets of the order of the real ones, as shown in Figure 8. Furthermore, a single block was placed across the wall thickness, unlike the two-leaf walls of the test structure. The model had about 200 brick-shaped blocks, which were assumed to behave as rigid bodies, each with 6 degrees-of-freedom.

Figure 8. Simplified 3DEC model of the stone structure.

4.3. Material Properties

Consistent with the aim of applying the simplest modeling assumptions, the joints were assigned the standard Mohr–Coulomb model with brittle behavior available in 3DEC [15]. Joint deformability is characterized by normal and shear stiffnesses, and joint strength by the friction angle, cohesion, and tensile strength. If either tensile or shear failure take place, the cohesion and tensile strengths are both set to zero, while the friction angle is unchanged.

In a rigid block model, all the deformation is concentrated at the joints, so the joint stiffness parameters have to be selected to reproduce the masonry elastic moduli. For the blind prediction before the test, a Young's modulus of 2077 MPa given by wallette tests was the available information. The estimation of the normal and shear joint stiffnesses was based on an average spacing of 0.3 m for horizontal joints and 0.5 m for the vertical ones [23]. However, for the post-diction analyses, the natural frequencies measured experimentally were provided. In order to match these, the joint stiffnesses had to be reduced by a factor of 2, implying that the tested masonry walls were softer than the wallettes.

The initial joint strength parameters in the prediction runs were based on the diagonal compression of the wallettes. Assuming a friction angle of 30°, a cohesion of 0.32 MPa was estimated, and the tensile strength was taken as half of the cohesion. In the post-test analyses, the same cohesive and tensile strengths were used, but the friction angle was increased to 35°, in order to obtain a better fit of the experimental response. The higher friction possibly accounted for the joint irregularity joints in the physical model. For the base joint, between model and shaking table, where no slip occurred, a friction angle of 45° was adopted.

4.4. Pushover and Dynamic Analyses

Prior to the dynamic runs, a quasi-static analysis was performed with the DE model to obtain the failure load in pushover-type tests. First, the gravity load was applied, with the base block fixed in all directions. To perform the pushover analysis, a static horizontal load, proportional to the block weight, was applied to all the blocks, in the out-of-plane direction of the façade. The horizontal load was increased in increments corresponding to a mass force of 0.05 g, until collapse. The failure mode, shown in Figure 9a, agreed quite well with the deformation pattern observed experimentally, involving the out-of-plane rotation of the façade wall, while the side wall with the opening broke into a four component mechanism (Figure 7c). The pushover collapse load was 0.65 g, well below the 1.05 g record that led the model to near-collapse in the shake table tests. In any case, the rather simplified geometry of the DE rigid block structure appeared to be able to represent the key features of the experimental behavior.

(a) (b)

Figure 9. Numerical model configuration: (**a**) pushover failure mechanism; (**b**) final configuration at end of dynamic analysis.

For the dynamic runs, the model was first brought to equilibrium under the gravity load. Then, a time domain dynamic analysis was performed. The model was loaded by applying at the base block the dynamic records measured in the shake table during the tests. Rayleigh damping was employed, including both the mass- and stiffness-proportional components, with a value of 2% of critical damping at the fundamental frequency of 10 Hz. This value was selected to improve the simulation of the final test with the strongest motion, which caused widespread damage. Thus, in the early test stages, with lower input, the numerical model was expected to display higher response peaks than the experiment. It should be noted that in the DE model, further energy dissipation took place by friction on the sliding joints. The typical time step in the dynamic runs, determined by the code for reasons of numerical stability, was in the order of 6×10^{-6} s. The use of the stiffness-proportional component of Rayleigh damping was mostly responsible for such a small value of the time step.

The model configuration after the last dynamic stage, with a peak acceleration of 1.05 g, is depicted in Figure 9b, showing a clear resemblance with the observed behavior (Figure 7c). The peak displacements at the monitored points of the structure were also generally in reasonable agreement with the experiment [23]. Naturally, we cannot expect to match the details of the time response curves.

In conclusion, the reported analysis was based on a rather simplified block model. It is certainly possible to improve the geometrical resemblance of the 3DEC representation, namely including the two-leaf wall structure, or the horizontal joint non-planarity, as discussed above. In addition, the simple joint constitutive model adopted does not allow the representation of the progressive accumulation of damage in the successive tests. It should be noted, however, that data to support those refinements is not usually available. In addition, it was found that the material deformability provided by the static tests of elementary block assemblages did not yield the natural frequencies measured experimentally. Therefore, the calibration of the numerical model with the actual structural frequencies observed was an essential step to improve its performance in the dynamic runs.

5. Discussion

Discrete element models are a flexible analysis tool for discontinuous media that can be exploited in many different ways. During model generation, various levels of detail can be introduced, so that a better resemblance to the real block structure can be progressively achieved by introducing finer detailing. The new technologies that record accurately the wall morphology make it possible nowadays. When these are not available, generation methods, such as discussed above, can be used to create numerical models that match the main patterns and features of the particular type of masonry under study. The results of the research project reported in the previous section indicate that a relatively simple block pattern, which did not attempt to reproduce each individual block, was able to provide a fairly good agreement with the experimental data. More data is required for a refined model, and higher computational costs are to be expected. Therefore, the need to simplify is obvious. More research effort needs to be devoted to evaluating the influence of model simplification on the numerical results, providing guidance for engineering practice.

Pushover analysis provides a powerful tool for seismic assessment. In DE models, the physically possible failure mechanisms can be viewed and evaluated. The effect of material properties can be checked by parametric studies, or by using a statistical variation of parameters, with reasonable computational costs. Time domain dynamic analysis is much more computationally intensive. The explicit algorithms employed in most DE codes allow a rigorous simulation of the evolving geometry and contact conditions, but typically require very small time-steps. A key requirement for a meaningful dynamic analysis of existing structures is a good simulation of the natural frequencies, as several studies have shown [17,23]. In situ measurements appear, therefore, highly desirable to guarantee that the numerical model matches the dynamic characteristics of the real structure.

Shake table tests of masonry typically display a significant variability in the results. This has been observed even for rocking of a single stone block, for which successive tests with the same earthquake record sometimes varied significantly [26]. For a column of marble drums, a relatively wide range of responses was also obtained [27]. Large amplitude block rocking is a phenomenon known to be sensitive to the initial conditions and loading. The change in local contact conditions between blocks in successive tests, whenever slip takes place, can account for part of these variations. DE models with slightly different system geometry, properties, or input loads, have also been shown to lead to a clear dispersion of the results [3]. From a practical point of view, this sensitivity implies that a single dynamic analysis is not sufficient, and thus multiple runs, varying the main input parameters of the model, are always necessary to reach reliable conclusions.

6. Concluding Remarks

Discrete element models are an important tool in the analysis of masonry, given their ability to reproduce the observed patterns of behavior, particularly failure modes defined by the block structure. Analysis of masonry structures under intense earthquake loading has been one of the main areas of application of these models. For a successful application in engineering practice, any numerical tool needs to prove accurate and robust. Validation studies involving comparisons with shake table tests or the observed response during earthquakes are fundamental to provide confidence in a numerical method for seismic assessment studies. Complex models, however, often require data that is not readily available for existing masonry constructions. Model simplification strategies, whether in the representation of the block geometry or in the constitutive assumptions, are a key to providing meaningful results with the existing data in practical situations. The improvement of modeling procedures remains a major research goal.

Conflicts of Interest: The authors declare no conflict of interest.

References

1. Lourenço, P.B. Computations of historical masonry constructions. *Prog. Struct. Eng. Mater.* **2002**, *4*, 301–319. [CrossRef]
2. Lemos, J.V. Discrete element modeling of masonry structures. *Int. J. Archit. Herit.* **2007**, *1*, 190–213. [CrossRef]
3. Psycharis, I.N.; Lemos, J.V.; Papastamatiou, D.Y.; Zambas, C.; Papantonopoulos, C. Numerical study of the seismic behaviour of a part of the Parthenon Pronaos. *Earthq. Eng. Struct. Dyn.* **2003**, *32*, 2063–2084. [CrossRef]
4. Mendes, N.; Zanotti, S.; Lemos, J.V. Seismic performance of historical buildings based on discrete element method: An adobe church. *J. Earthq. Eng.* **2018**. [CrossRef]
5. Cakti, E.; Saygili, O.; Lemos, J.V.; Oliveira, C.S. Discrete element modeling of a scaled masonry structure and its validation. *Eng. Struct.* **2016**, *126*, 224–236. [CrossRef]
6. Godio, M.; Beyer, K. Evaluation of force-based and displacement-based out-of-plane seismic assessment methods for unreinforced masonry walls through refined model simulations. *Earthq. Eng. Struct. Dyn.* **2018**, 1–22. [CrossRef]
7. Galvez, F.; Sorrentino, L.; Ingham, J.; Dizhur, D. One way bending capacity prediction of unreinforced masonry walls with varying cross section configurations. In Proceedings of the 10th International Masonry Conference, Milan, Italy, July 2018; Milani, G., Taliercio, A., Garrity, S., Eds.; The International Masonry Society: Whyteleafe, UK; pp. 641–653.
8. Malomo, D.; DeJong, M.J.; Penna, A. Distinct element modelling of the in-plane failure mechanisms of URM walls. In Proceedings of the 10th International Masonry Conference, Milan, Italy, July 2018; Milani, G., Taliercio, A., Garrity, S., Eds.; The International Masonry Society: Whyteleafe, UK; pp. 581–594.
9. Casolo, S.; Uva, G. Nonlinear analysis of out-of-plane masonry façades: full dynamic versus pushover methods by rigid body and spring model. *Earthq. Eng. Struct. Dyn.* **2018**, *42*, 499–521. [CrossRef]
10. Malomo, D.; Pinho, R.; Penna, A. Using the applied element method for modelling calcium silicate brick masonry subjected to in-plane cyclic loading. *Earthq. Eng. Struct. Dyn.* **2018**, *47*, 1610–1630. [CrossRef]
11. Garofano, A.; Lestuzzi, P. Seismic Assessment of a Historical Masonry Building in Switzerland: The "Ancien Hôpital De Sion". *Int. J. Archit. Herit.* **2016**, *10*, 975–992. [CrossRef]
12. Estêvão, J.M.C.; Oliveira, C.S. A new analysis method for structural failure evaluation. *Eng. Fail. Anal.* **2015**, *56*, 573–584. [CrossRef]
13. Milani, G.; Lourenço, P.B. 3D non-linear behavior of masonry arch bridges. *Comput. Struct.* **2012**, *110–111*, 133–150. [CrossRef]

14. Chiozzi, A.; Milani, G.; Grillanda, N.; Tralli, A. A fast and general upper-bound limit analysis approach for out-of-plane loaded masonry walls. *Meccanica* **2018**, *53*, 1875–1898. [CrossRef]

15. Itasca. *3DEC—Three-dimensional Distinct Element Code*; Version 5.20; Itasca Consulting Group: Minneapolis, MN, USA, 2017.

16. Sarhosis, V.; Sheng, Y. Identification of material parameters for low bond strength masonry. *Eng. Struct.* **2014**, *60*, 100–110. [CrossRef]

17. Lemos, J.V.; Oliveira, C.S.; Navarro, M. 3D nonlinear behavior of an obelisk subjected to the Lorca May 11, 2011 strong motion record. *Eng. Fail. Anal.* **2015**, *58*, 212–228. [CrossRef]

18. Mordanova, A.; De Felice, G. Seismic assessment of archaeological heritage using discrete element method. *Int. J. Archit. Herit.* **2018**. [CrossRef]

19. De Felice, G. Out-of-plane seismic capacity of masonry depending on wall section morphology. *Int. J. Archit. Herit.* **2011**, *5*, 466–482. [CrossRef]

20. Lemos, J.V.; Costa, A.C.; Bretas, E.M. Assessment of the seismic capacity of stone masonry walls with block models. In *Computational Methods in Earthquake Engineering*; Papadrakakis, M., Fragiadakis, M., Lagaros, N.D., Eds.; Springer: Dordrecht, Netherlands, 2011; pp. 221–235.

21. Candeias, P.X.; Costa, A.C.; Mendes, N.; Costa, A.A.; Lourenço, P.B. Experimental assessment of the -of-plane performance of masonry buildings through shaking table tests. *Int. J. Archit. Herit.* **2017**, *11*, 31–58. [CrossRef]

22. Mendes, N.; Costa, A.A.; Lourenço, P.B.; Bento, R.; Beyer, K.; Felice, G.; Gams, M.; Griffith, M.; Ingham, J.; Lagomarsino, S.; et al. Methods and approaches for blind test predictions of out-of-plane behavior of masonry walls: A numerical comparative study. *Int. J. Archit. Herit.* **2017**, *11*, 59–71.

23. Lemos, J.V.; Campos Costa, A. Simulation of shake table tests on out-of-plane masonry buildings. Part (V): Discrete element approach. *Int. J. Archit. Herit.* **2017**, *11*, 117–124. [CrossRef]

24. AlShawa, O.; Sorrentino, L.; Liberatore, D. Simulation of shake table tests on out-of-plane masonry buildings. Part (II): Combined finite-discrete elements. *Int. J. Archit. Herit.* **2017**, *11*, 79–83. [CrossRef]

25. Cannizzaro, F.; Lourenço, P.B. Simulation of shake table tests on out-of-plane masonry buildings. Part (VI): Discrete element approach. *Int. J. Archit. Herit.* **2017**, *11*, 125–142. [CrossRef]

26. Peña, F.; Prieto, F.; Lourenço, P.B.; Campos Costa, A.; Lemos, J.V. On the dynamics of rocking motion of single rigid-block structures. *Earthq. Eng. Struct. Dyn.* **2017**, *36*, 2383–2399. [CrossRef]

27. Papantonopoulos, C.; Psycharis, I.N.; Papastamatiou, D.Y.; Lemos, J.V.; Mouzakis, H. Numerical prediction of the earthquake response of classical columns using the distinct element method. *Earthq. Eng Struct. Dyn.* **2002**, *31*, 1699–1717. [CrossRef]

buildings

MDPI

Article

Diagnosis and Seismic Behavior Evaluation of the Church of São Miguel de Refojos (Portugal)

Rafael Ramírez *, Nuno Mendes and Paulo B. Lourenço

ISISE, Institute of Science and Innovation for Bio-Sustainability (IB-S), 4800-058 Guimarães, Portugal;
nunomendes@civil.uminho.pt (N.M.); pbl@civil.uminho.pt (P.B.L.)
* Correspondence: rafael.alvarezdelara@gmail.com; Tel.: +351-253-510-200

Received: 7 May 2019; Accepted: 28 May 2019; Published: 31 May 2019

Abstract: The Benedictine Monastery of São Miguel de Refojos, located in Cabeceiras de Basto (Portugal), is a monumental complex and a distinctive example of the 18th century Portuguese Baroque architecture. This study addresses the state of conservation of the church as well as the evaluation of its structural behavior and seismic performance. An initial inspection and diagnosis campaign revealed that the structure presents low to moderate damage and other non-structural issues generally associated with high levels of moisture and water infiltration. In order to study the structural performance, a three-dimensional (3D) numerical model was prepared based on the finite element method. This model was calibrated with respect to dynamic identification tests and nonlinear static analyses were then performed to evaluate the seismic behavior. Capacity curves, deformations, crack patterns, and failure mechanisms were used to characterize the structural response. Additionally, the safety evaluation for horizontal actions was verified by means of limit analysis. An overall good agreement was found between the results of the pushover and the limit analyses. To conclude, the present work provides a comprehensive evaluation of the state of conservation of the church and verifies the safety condition of the structure for seismic actions.

Keywords: unreinforced masonry structure; finite element modelling; nonlinear static analysis; limit analysis; seismic behavior; safety assessment

1. Introduction

The church of São Miguel de Refojos is a monumental building in Baroque-Rococo style and is part of the historical Benedictine Monastery located in Cabeceiras de Basto, Portugal. The monastery was founded in the 12th century by the Benedictine Order and was located in an unpopulated but geographically strategic area [1]. After some flourishing early years, the building underwent a ruinous period during the 15th century. However, by the end of the following century, the monastery experienced a new economic and architectural recovery with the change of administration [1,2]. Major works for the renovation and re-organization of the monastic complex started during the last decades of the 17th century [3]. In fact, the current aspect of the church is the result of the works carried out between 1675 and 1766, when it was renovated in Baroque style [4].

The suppression of the Religious Orders in 1834 led to a progressive abandonment of the Monastery and the building started a new period of decay that lasted until the first half of the 20th century [5]. After that, the building was subjected once more to different restoration and repair works, namely filling of cracks, cleaning of stone surfaces, repainting, application of water drainage collectors, reconstruction of the roof and replacement of tiles. Nowadays, the monastery is used by the municipal authorities of Cabeceiras de Basto and it still maintains religious activities. Moreover, the church and the sacristy of the monastery have been classified as a Public Interest Building since 1933.

The church of São Miguel de Refojos stands as a historical and cultural landmark and possesses an unquestionable artistic and architectural value. However, as for many other historical structures, environmental and accidental actions always entail a major risk that may lead to degradation, severe damage, or even loss. In particular, unreinforced masonry (URM) buildings, such as the church of São Miguel, are vulnerable to the horizontal loads caused by earthquakes [6,7]. The seismic performance of URM structures is mainly determined by: the material properties of masonry (high-specific mass, low tensile and shear strength, brittle response), the geometrical configuration, the mass and stiffness distribution, and the type and quality of the connections between the different structural components [8]. In addition, the global response of this type of construction is difficult to characterize since URM structures dissipate energy by the propagation of damage (cracks), which consequently leads to isolated mechanisms and local failure modes [9].

In this context, finite element (FE) modelling is a very useful tool for URM cases because it allows for an accurate simulation of the overall structural behavior under different scenarios. Therefore, numerical approaches may be used to assess the safety condition of historical buildings as well as to evaluate the need and efficiency of different retrofitting techniques. In addition, the case study presented in this paper presents a comprehensive methodology (inspection and diagnosis, numerical modelling, structural analyses, and safety assessment) that can be extrapolated and applied to similar constructions.

2. Description of the Church

The church of São Miguel de Refojos is located on the southern side of the monastery. It has a Latin-cross plan (nave, transept, and chancel) and follows a classical east–west orientation (Figure 1). In the plan, the building is about 60 m long and 24 m wide in the transept. At one and the other side of the nave and chancel, adjacent spaces with smaller dimensions are used to accommodate different ecclesiastical functions. The transept has a direct connection with the central cloister on the left side of the church and it also connects with an octagonal plan chapel located outside the main body of the church. The detached position of this volume seems to indicate that it was built in a later period [2].

The front facade has two bell towers with square plan and of about 40 m high. Inside the church, the main entrance is covered by a choir loft supported by a ribbed vault. The crossing is covered by a dome with an oval-shaped plan standing over pendentives. In turn, the body of the dome is made up by a drum, a cupola, and a lantern. The interior height of the church is about 18 m in the nave, 35 m under the dome, and 16 m in the chancel.

The ceilings of the nave, transept, and chancel are made up by barrel vaults. The spaces above the vaults are covered by gable roofs with ceramic tiles finishing, supported by timber trusses oriented in the shortest span. Due to access limitations, only the extrados of the vault and the roof structure of the nave can be inspected.

3. Inspection Works and Diagnosis

A comprehensive inspection and diagnosis campaign were carried out in order to characterize the main structural aspects and evaluate the damage condition and state of conservation of the church. The works included: geometrical and damage survey, monitoring of temperature and relative humidity inside the church, dynamic identification tests, georadar tests, and several geotechnical works.

3.1. Geometrical and Damage Survey

The identification of damage involved a detailed visual inspection of all of the structural elements, both from the outside and the inside of the building. The inspection revealed the presence of structural and non-structural problems. The non-structural damage corresponds to moisture stains, vegetation, fungi, and deterioration of plaster and stone. Additionally, the building shows low to moderate structural damage. There are cracks at the top of the south facade, on the transept walls, on the walls behind the high altar, and on the arch between the crossing and the chancel. The most significant damage appears in the connections between the front facade and the south tower, and in between this tower and the south wall of the nave. Moreover, the open cracks allow for rainwater infiltration with the subsequent deterioration of the internal elements. The floors of the high altar and the octagonal chapel present significant deformations probably caused by consolidation of the soil. Similarly, other deformations in the structure might be associated with a foundation settlement.

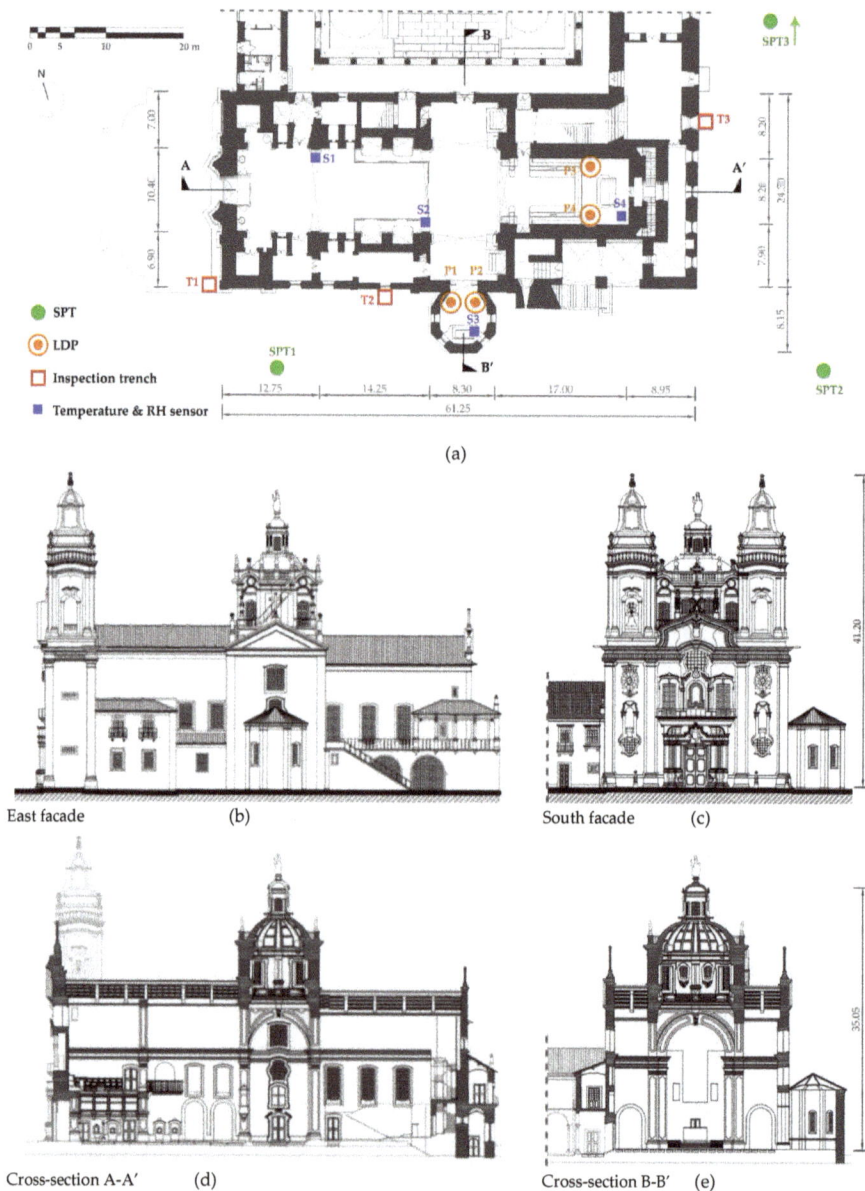

Figure 1. Architectural definition of the church of São Miguel de Refojos: (**a**) ground floor plan, (**b**) longitudinal elevation, (**c**) front elevation, (**d**) longitudinal cross-section, (**e**) transversal cross-section. Dimensions in m.

3.2. Monitoring of Temperature and Relative Humidity

Air temperature and relative humidity are generally associated with mechanical stresses due to thermal expansion and hygroscopic swelling, opening and closing of cracks, deformations, and other non-structural problems, such as condensation of water vapor on cold surfaces. Temperature, relative humidity, and dew point were monitored for twelve months in four different locations inside the church (Figure 1a). The average values of the temperature and relative humidity measurements is represented in Figure 2.

Considering the entire monitoring period for the four locations, the average total, maximum and minimum temperatures are 15 °C, 27 °C, and 4 °C, respectively. On the other hand, the overall average and average maximum and minimum relative humidity values inside the church are equal to 71%, 97%, and 35%, respectively. This represents a yearly average thermal excursion of 23 °C and a yearly average humidity variation equal to 62%. Nevertheless, the daily variation of temperature and relative humidity values inside the church is less significant.

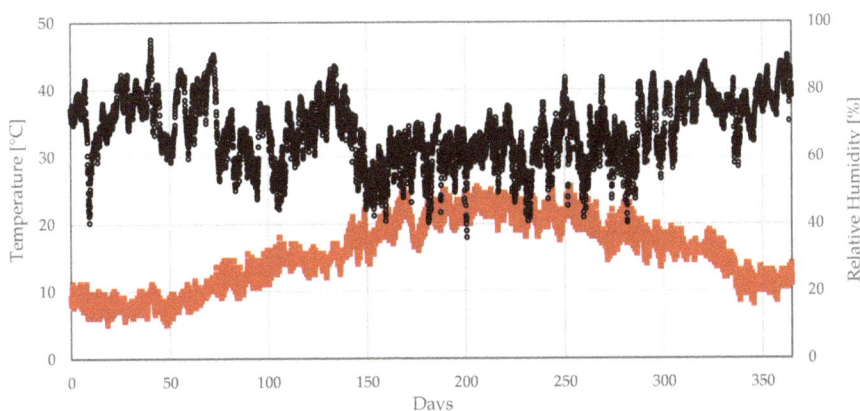

Figure 2. Average temperature and relative humidity values measured inside the church.

3.3. Dynamic Identification Tests

Dynamic identification tests were carried out in order to estimate the dynamic properties of the church (natural frequencies and mode shapes), which would be used later to validate the numerical model. The accelerations of the structure caused by ambient vibrations (wind, traffic, etc.) were recorded using piezoelectric accelerometers with 10 V/g sensitivity (±0.5 g pk), acquisition boards with 24-bit resolution, and a USB chassis connected to a computer. The acceleration time series were acquired with a sampling frequency equal to 200 Hz. The total duration of the signals was 30 min.

The vibrations were measured at the top of the walls of the nave, dome, and chancel, in both the transversal (X) and longitudinal (Y) directions of the church, for five different configurations (Figure 3). The setups were correlated using a reference accelerometer placed on the balcony of the dome in the transversal direction of the church. The chosen reference point was considered sufficient for the whole set of setups since the transversal direction was expected to be more flexible and therefore more representative than the longitudinal one.

The signs of the dynamic identification tests were processed in ARTeMIS Modal software [10] by using two different methods, namely: (a) Enhanced Frequency Domain Decomposition (EFDD) and (b) Stochastic Subspace Identification based on Unweighted Principal Component (SSI-UPC). Both methods allowed us to estimate six modes, with frequencies ranging from 1.79 Hz to 5.14 Hz. The six estimated modes are associated with local and global modes mainly in the transversal direction, as expected.

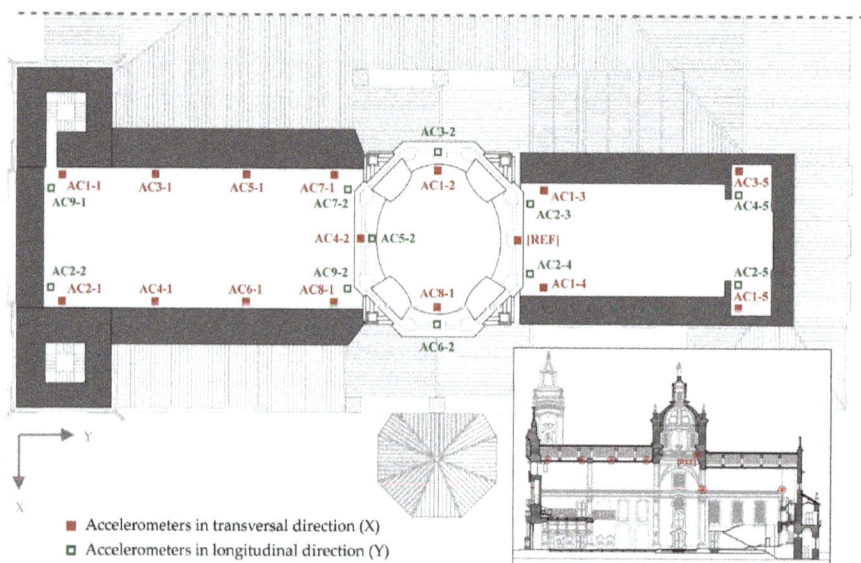

Figure 3. Location of the accelerometers for the dynamic identification tests.

The vibration modes obtained by the two aforementioned methods were compared using the MAC (Model Assurance Criterion) [11]. In a range from 0 to 1, where 1 indicates a perfect match of the mode shapes, the MAC of the first three modes of vibration show values between 0.89 and 0.99, which indicates that the results of the two methods are very close. However, the remaining three modes associated with the higher frequencies present low MAC values, with an average equal to 0.56.

Therefore, only the first three modes are of interest for the scope of this study (Figure 4). The first mode of vibration (1.79 Hz) is associated with a local mode of the nave related to the vibration of the towers in the transversal direction. The second mode (2.75 Hz) is the first global mode, with simple curvature in the transversal direction of the church. Finally, the third vibration mode (3.72 Hz) presents double curvature, corresponding to a global translation of the church in the transversal direction.

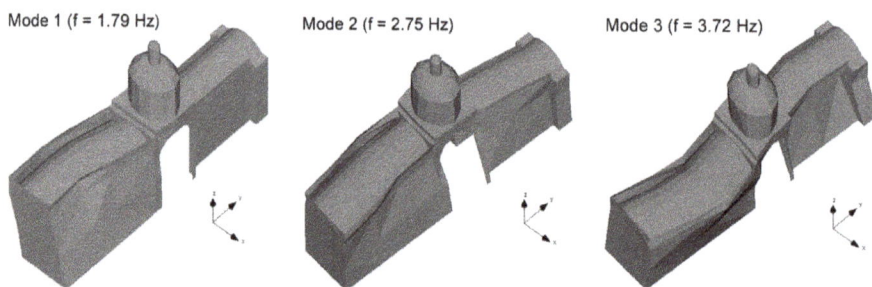

Figure 4. First three modes of vibration estimated by the Stochastic Subspace Identification based on Unweighted Principal Component (SSI-UPC) method.

3.4. Georadar Tests

Ground Penetrating Radar (GPR) tests were performed in different parts of the monastery using equipment with 500, 800, and 1600 MHz antennas. These tests were aimed at identifying the unseen constructive elements and characterizing the material composition of the structure. The use of Ground Penetrating Radar on historical structures has already been documented by different authors [12,13].

In some cases, the constitution of the walls is visible at plain sight because there is no rendering layer covering the surface. That is the case of the thinner walls behind the altar, made up by a unique leaf of ashlar masonry. For the rest of the walls, the Ground Penetrating Radar was used to identify the internal composition. Thus, GPR readings were carried out, both in the horizontal and in the vertical direction, in various locations of the walls, namely the main facade, the longitudinal walls of the nave, the walls of the transept, and the walls of the octagonal chapel. In general, the results show that the structure is made up by composite walls with three masonry leaves (Figure 5): an outer layer of irregular stone blocks, a similar internal layer, and a rubble core.

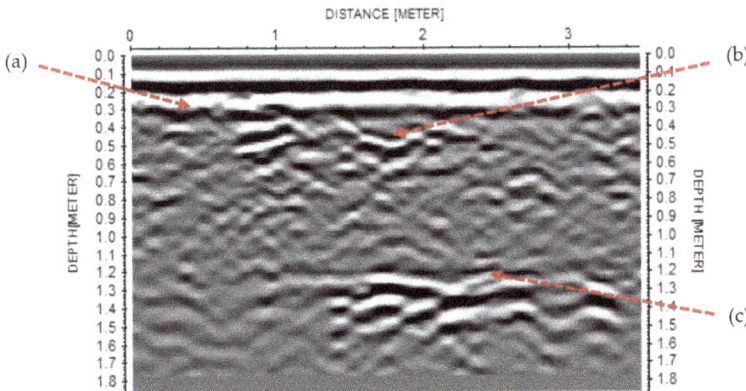

Figure 5. Horizontal georadar reading on the south wall of the nave: (**a**) inner surface of the interior leaf at 35–40 cm depth; (**b**) inner surface of the interior leaf at 55 cm depth (irregular inner surface); (**c**) opposite side (inner surface of the exterior leaf) at 1.20–1.25 m depth.

In addition, georadar readings were used to identify the constitution of the barrel vault of the nave. The radargrams showed that the total thickness of the vault is about 16 cm. Furthermore, the readings next to the springing line showed an irregular infill, which was characterized as a mixture of gravel and waste.

3.5. Geotechnical Survey

The geotechnical prospection plan consisted of three Standard Penetration Tests (SPT) outside the church and four Light Dynamic Penetrometer tests (LDP) inside the church, as well as several inspection trenches at the base of the walls and water table measurements, see Figure 1a. The purpose of the geotechnical prospection was manifold: characterization of the soil profile and its mechanical properties, evaluation of the type of foundation and identification of the foundation level, verification of the existence of a draining system, and detection of the water table depth.

The trenches and boreholes of the drilling tests showed that the water table is very close to the ground level of the church (summer period), less than 1.00 m depth in some cases (Figure 6). The shallow water level made it impossible to identify the actual depth of the foundations under the walls. Nonetheless, the geotechnical survey allowed us to identify the geological profile. The superficial layer is constituted by soft clay, with low resistance and very deformable, until a depth of about 4.00 m. The following layer corresponds to a transition layer of gravelly sand. Finally, the last layer of the soil consists of highly altered and fractured rock (starting at depths between 4.20 to 5.50 m).

The existence of a surface layer with poor geotechnical characteristics might be the cause of the settlements detected on the church pavements. Some of the observed structural damages could also be the result of localized foundation settlements.

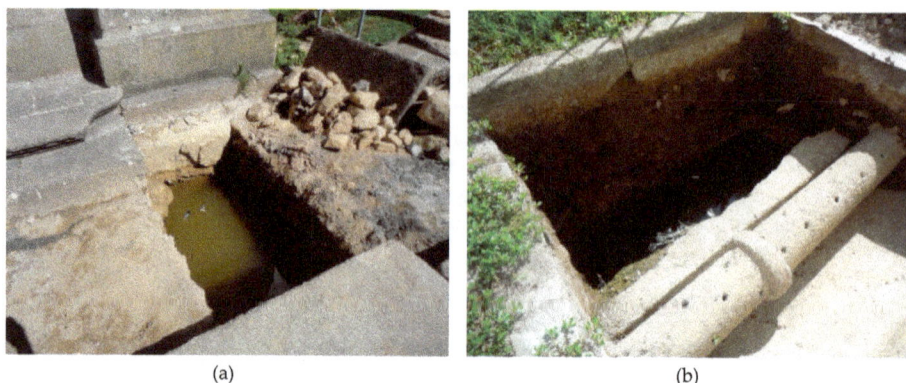

(a)   (b)

Figure 6. Water table visible in the inspection trenches days after the excavation (summer period). (a) Excavation 1: water level at −0.93 m below the nave floor; (b) Excavation 3: water table at −1.10 m measured from the floor of the nave.

4. Preparation and Calibration of the Numerical Model

The numerical Finite Element (FE) model of the church was prepared using the structural analysis software DIANA (DIsplacement ANAlyser) [14].

4.1. Geometry

The geometry of the three-dimensional (3D) model included all major structural elements, namely the main volume of the church plus the annexed lateral bodies (Figure 7). The stiffness of the structural elements of the monastery adjacent to the church (orthogonal walls and columns of the cloister) was considered in the model by means of spring elements. Since it was not possible to inspect the roofs of the transept and the chancel, the truss elements in these areas were idealized according to the known roof structure above the nave.

4.2. Type of Elements

Different elements were used to define the mesh: four-node tetrahedron solid elements (walls, arches and columns), three-node curved shell elements (vaults), three-node three-dimensional class-III beam elements (roof trusses), and spring elements (stiffness of adjacent buildings). For later nonlinear analyses, in addition, the number of integration points across the thickness of the shell elements was increased (five integration points).

4.3. Boundary Conditions

An overall base level located at −2.00 m below the nave floor was assumed for all of the walls. At that level, the foundation was considered rigid. Thus, all degrees of freedom of the nodes at the base of the masonry walls were constrained. In addition, as already mentioned, spring elements were introduced to simulate the stiffness of the adjacent structures of the monastery.

4.4. Loads

Besides the self-weight of the modelled structural elements, which is automatically calculated by the software, the weight of the roof and the infill material of the vaults was additionally applied. The weight of the roof was assumed equal to 1.00 kN/m^2 and it was applied as a uniformly distributed load along the trusses. The infill material was considered up to half the height of the vaults [15] and its weight was estimated equal to 12 kN/m^3. The infill was modelled as a distributed superficial load applied to the extrados of the vaults, near the supports.

Figure 7. Plan view with the geometry considered for the numerical model.

4.5. Material Properties

As it was mentioned before, the visual inspection and georadar tests allowed us to identify the different types of materials that compose the structure of the church, namely brick masonry (BRM–vaults), ashlar masonry (GRM–walls behind the high altar, arches in the crossing, top part of the towers, and dome), three-leaf stone masonry (STM–main walls and rest of masonry elements), granite units (lintels), and timber (roof truss structure). Figure 8 shows the material configuration considered in the model.

In the first stage, the material properties were defined according to sets of data and recommendations available in specialized literature. In particular, the prescriptions of the Italian code [16,17] were used to determine the elastic properties of the masonry materials: average density, elastic modulus, and Poisson's ratio calculated from the range of values proposed for each type of masonry. Similarly, the characteristics of granite and timber elements were defined as the mean value of the material properties provided by specialized studies [18,19].

The initial linear-elastic properties were subsequently updated during the calibration process of the numerical model. This numerical validation was done with respect to the experimental dynamic properties estimated from the dynamic identification tests, namely the frequencies of the first three modes of vibration. The calibration was performed following the Douglas-Reid method [20], which involves an optimization process to minimize the difference between the experimental and numerical responses. In particular, the elastic moduli of the three masonry materials (BRM, GRM, and STM) were used as the variables to calibrate. The upper and lower limits for the variables were defined in accordance with the aforementioned Italian code, assuming the highest and lowest values of the range proposed for each masonry material.

After the optimization process, the numerical model was able to simulate the first three experimental modes of the church with an average error of the frequency values less than 2%. Moreover, the mode shapes of the numerical model were compared with the modal configurations of the experimental modes through MAC (Modal Assurance Criterion) values [11]. From 0 to 1, where 1 indicates a perfect match, the calibrated model of the church resulted in an average MAC value equal to 0.92. The numerical model was thus successfully validated, and it was possible to use the updated material properties for further analyses.

Figure 8. Geometry of the numerical model with identification of materials.

Using the updated linear-elastic parameters, the necessary nonlinear properties were estimated following relations established in literature or adopting values coming from similar case studies. In particular, the nonlinear behavior of the masonry materials (BRM, GRM, and STM) was described using the Total Strain Rotating Crack model, which corresponds to a smeared crack model based on the principal strains [14].

An exponential function was considered for the tensile softening and a parabolic curve was assumed for compression (Figure 9). In this material model, the shear behavior is updated according to the damage which occurred during the analysis. The compressive strength (f_c) was determined as a function of the modulus of elasticity (E), where $f_c = E/600$ [21]. In turn, the compression fracture energy was determined as a function of the compressive strength and the ductility index ($d_{u,c}$), where $d_{u,c} = G_c/f_c$ and $d_{u,c} = 1.6$ mm [22]. The tensile strength and tensile fracture energy of masonry were considered equal to 0.15 MPa and 50 N/m, respectively [23].

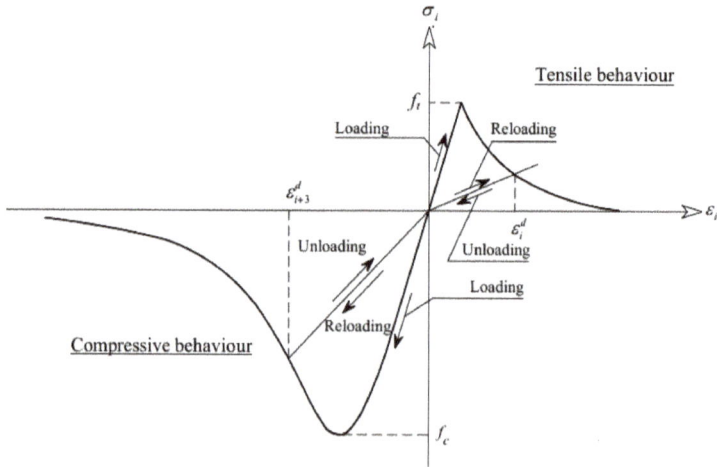

Figure 9. Nonlinear behavior of the masonry materials in the numerical model: Total Strain Rotating Crack model [24].

Table 1 presents the updated linear and nonlinear properties of the materials used in the model. It must be noted that the granite and timber elements were assumed to work always within their linear-elastic range. Finally, the crack bandwidth value (h) was automatically calculated by the software according to the type of element, namely $h = \sqrt{2A}$ and $h = \sqrt{3V}$ for shell and solid elements respectively, where A and V stand for the area and volume of the element.

Table 1. Linear-elastic and nonlinear properties of materials.

Property	Symbol	Unit	Material				
			Brick Masonry (BRM)	Three-Leaf Stone Masonry (STM)	Ashlar Masonry (GRM)	Granite	Timber
Density	ρ	kg/m^3	1800	2000	2100	2500	600
Elastic modulus	E	MPa	1360	1620	1900	30,000	13,000
Poisson's ratio	ν	-	0.20	0.20	0.20	0.20	0.30
Compressive strength	f_c	kPa	2260	2700	3160	-	-
Compressive fracture energy	G_c	kN/m	3.60	4.30	5.05	-	-
Tensile strength	f_t	kPa	150	150	150	-	-
Tensile fracture energy	G_f^I	kN/m	0.05	0.05	0.05	-	-

5. Seismic Behavior Evaluation

5.1. Nonlinear Static Analysis

The structural performance of the church for the seismic action was initially verified using nonlinear static or pushover analysis. The use of pushover analysis for the seismic assessment of historical masonry structures has been proven to be a suitable approach and is fairly documented in the literature [8,25,26]. In the current study, the nonlinear static analyses were carried out using a uniform unidirectional load pattern based on horizontal forces proportional to the mass of the structure at all elevations [27].

This type of analysis allows for the evaluation of the nonlinear behavior by applying the action in successive load steps. The equilibrium of the system of equations for each step is guaranteed by an iterative method and a convergence criterion. In this case, the regular Newton–Raphson iterative method and an energy-based convergence criterion (10^{-3}) were used. In addition, the Line Search algorithm and arc-length method were used as well to improve the convergence and obtain the post-peak response.

The seismic performance of the building was evaluated according to the global horizontal axes, X and Y, which correspond to the transversal and longitudinal directions of the church, respectively. The results of the pushover analyses are shown in Figure 10 in terms of capacity curves (displacement against base shear factor), using the point at the top of the south tower as a control node to plot the displacements. The base shear factor corresponds to the ratio between the sum of the horizontal forces and the total self-weight of the structure. It should be noted that the analysis in the negative transversal direction (–XX) was neglected due to the presence of the adjacent buildings introduced in the model as spring elements, which can provide a larger capacity in that direction.

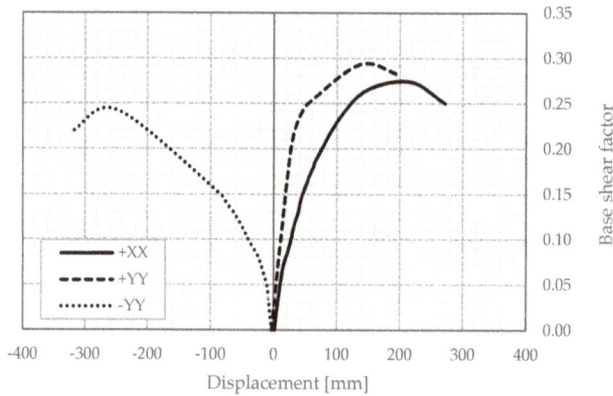

Figure 10. Capacity curves for nonlinear static analyses. Control point at the top of the south tower. The base shear factor corresponds to the ratio between the sum of the horizontal forces and the total self-weight of the structure [dimensionless].

5.1.1. Positive Transversal Direction (+XX)

The seismic analysis in the positive transversal direction (+XX) indicates that the maximum horizontal capacity is equal to 0.28 g (Figure 10). The horizontal maximum displacement occurs at the top of the south tower. The damage pattern is characterized by shear cracks in the plane of the walls and cracks in the connections between structural elements (Figure 11). The most severe damage occurs in the main facade (diagonal cracks) and in the connections of the south tower with the orthogonal walls.

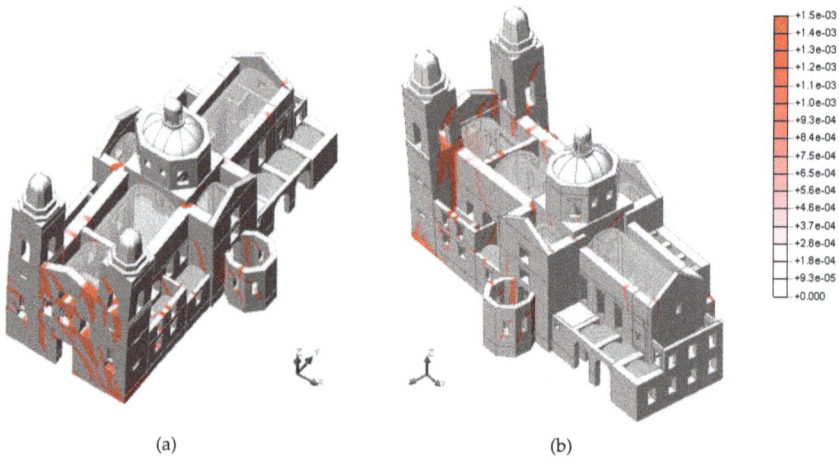

(a)   (b)

Figure 11. Crack width associated with the maximum principal strain at the end of the capacity curve for the pushover analysis in the positive transversal direction (+XX). (**a**) South-West view; (**b**) South-East view. Units in m.

5.1.2. Positive Longitudinal Direction (+YY)

The curve for the positive longitudinal direction shows a stiffer response, with a maximum capacity equal to 0.29 g (Figure 10). The maximum displacement occurs at the top of the towers. In fact, the main failure involves the local collapse of the belfries (Figure 12). In addition, there is also damage in the connections between the rear facade and the longitudinal walls of the chancel.

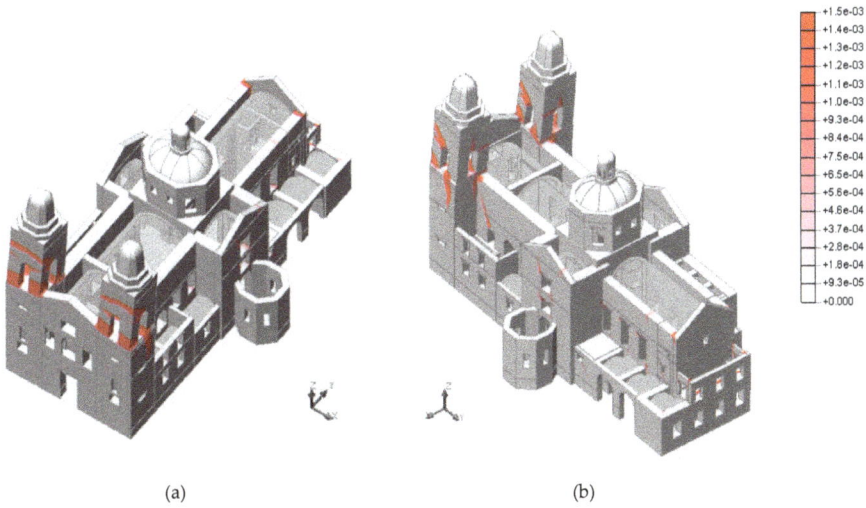

Figure 12. Crack width associated with the maximum principal strain at the end of the capacity curve for the pushover analysis in the positive longitudinal direction (+YY). (**a**) South-West view; (**b**) South-East view. Units in m.

5.1.3. Negative Longitudinal Direction (−YY)

The pushover analysis in the negative longitudinal direction (−YY) presents a maximum base shear factor equal to 0.24 (Figure 10). This value represents the lowest capacity obtained from the different pushover analyses and therefore indicates the most vulnerable direction of the church. This response is associated with a significant deformation of the main facade and the towers in the out-of-plane direction (Figure 13). The damage pattern is characterized by cracks in the connections between the towers and the longitudinal walls of the nave as well as vertical cracks in the main facade.

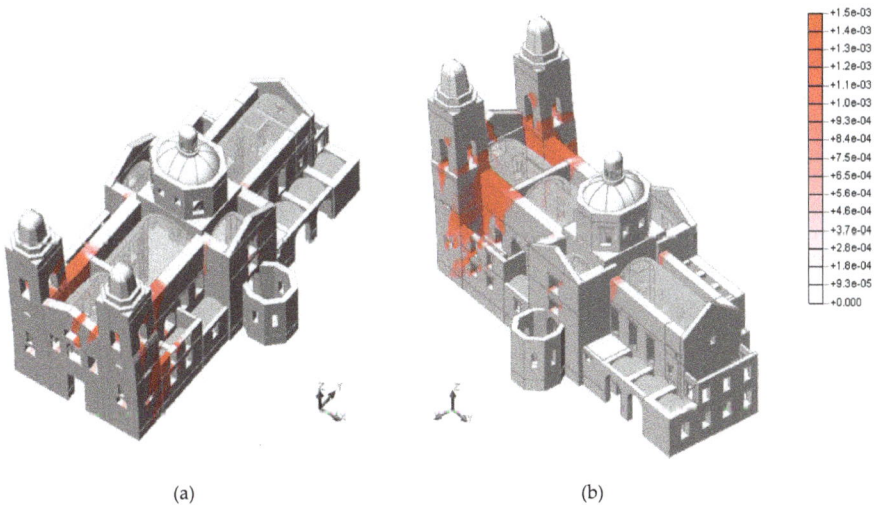

Figure 13. Crack width associated with the maximum principal strain at the end of the capacity curve for the pushover analysis in the negative longitudinal direction (−YY). (**a**) South-West view; (**b**) South-East view. Units in m.

5.2. Safety Assessment

The verification of the structural safety for the seismic action was carried out according to the Portuguese National Annex of Eurocode 8 [27]. The verification must be done for each type of seismic action, i.e., Type 1 (far field) and Type 2 (near field), in both principal (longitudinal and transversal) directions of the structure.

According to the Portuguese seismic zonation, the reference acceleration values (a_{gR}) for Cabeceiras de Basto are 0.05 g and 0.08 g for earthquakes Type 1 and Type 2, respectively. Moreover, the soil coefficient (S) is assumed equal to 1.0 (deep foundation on rock or other rock-like geological formation, including at most 5 m of weaker material at the surface) and the damping correction coefficient (η) is equal to 1.00 (5% damping).

The capacity required in terms of horizontal acceleration is determined as a function of a_{gR}, assuming collapse mechanisms based on macroblocks and defined taking into account the damage obtained from the numerical analyses. As presented in Table 2, the results of the pushover analyses for the seismic action indicate that the church meets the required criteria for both types of earthquake. The lowest capacity of the church occurs for the seismic action in the negative longitudinal direction (0.24 g) and is associated with a safety factor equal to 3.00. The safety factor for the seismic action is given by the relation between the base shear factor and the required capacity determined for the zone and defined in the Portuguese National Annex of Eurocode 8 (see Table 2).

Table 2. Safety verification for seismic action by means of nonlinear static analysis.

Direction	Maximum Capacity (g)	Required Capacity (g)	
		Type 1	Type 2
+XX	0.28	0.05	0.08
+YY	0.29	0.05	0.08
−YY	0.24	0.05	0.08

5.3. Limit Analysis

In order to validate the seismic performance of the church, the limit analysis based on the kinematic approach was also carried out. For the limit analysis, seven local collapse mechanisms were proposed taking into account the damage patterns caused in churches by past earthquakes [28], as well as the results from the pushover analyses and the existing damage in the structure. The stability for each mechanism was verified according to the methodology defined in the Italian code NTC-08 [16,17], in which the following criteria should be evaluated for the Ultimate Limit State:

(a) Verification A – Mechanisms involving part of the structure in contact with the soil:

$$a_0^* \geq \frac{a_g \cdot S}{q} \tag{1}$$

(b) Verification B – Mechanisms involving part of the structure above the ground level:

$$a_0^* \geq \frac{S_e(T_1) \cdot \psi(Z) \cdot \gamma}{q} \tag{2}$$

In the previous equations, a_0^* corresponds to the spectral acceleration for activation of the mechanism, and q is the behavior coefficient (2.00). Additionally, in Equation (1), $a_g = a_{gR} \cdot \gamma_I$ represents the peak ground acceleration (PGA) according to the seismic zonation (0.05 g and 0.08 g for earthquake Type 1 and Type 2, respectively) amplified by the coefficient of importance (1.00), and S is the soil coefficient (1.0). For the second verification, $S_e(T_1)$ is the spectral acceleration associated with the period T_1 (first mode of vibration of the structure in the considered direction), $\psi(Z) = Z/H$ is the ratio between the height of the hinge at the base of the collapse block divided by the total height, and $\gamma = 3N/(2N+1)$ is an amplification factor that takes into account the number of floors (N).

The spectral acceleration for activation of the mechanism is determined by:

$$a_0^* \geq \frac{a_0 \cdot g}{e^* \cdot FC} \tag{3}$$

where g is the acceleration of gravity (9.81 m/s²), a_0 is the load factor that activates the mechanism (determined by the Principle of Virtual Works for the equilibrium between the overturning moment caused by the horizontal forces and the stabilizing moment caused by the vertical forces), FC is the confidence factor associated to the level of knowledge of the structure (1.00 for $LC = 3$), and e^* corresponds to the mass participation factor of the mechanism,

$$e^* = \frac{g \cdot M^*}{\sum_i P_i} \tag{4}$$

where M^* is the mass participation of the mechanism defined as:

$$M^* = \frac{\left(\sum_i P_i \delta_{x,i}\right)^2}{g \cdot \sum_i P_i \delta_{x,i}^2} \tag{5}$$

where P_i is the vertical force of the self-weight and $\delta_{x,i}$ is the horizontal virtual displacement of the center of gravity of the macroblock i.

In the studied cases, the masonry materials were assumed to have infinite compressive strength and tensile strength equal to zero. The following collapse mechanisms were considered (Figure 14):

- Mechanism 1: Overturning of the lantern with rotation at the base (Figure 14a);
- Mechanism 2: Overturning of the belfry with rotation at the base (Figure 14b);
- Mechanism 3: Overturning of the gable of the rear facade (high altar) with rotation at the base and diagonal wedges from the lateral walls of the chancel (Figure 14c);
- Mechanism 4: Overturning of the main facade and towers with rotation at the base (Figure 14d);
- Mechanism 5: Overturning of the main facade with rotation at the base (Figure 14e);
- Mechanism 6: Overturning of the south tower with rotation at the base in the direction perpendicular to the facade (Figure 14f);
- Mechanism 7: Overturning of the south tower with rotation at the base in the direction parallel to the facade (Figure 14g).

It must be noted that these collapse mechanisms were defined based on expert opinions and typical failures caused by earthquakes, which can be different from the failure modes obtained from the pushover analysis. The results of the limit analysis for the seismic action are shown in Table 3. The maximum capacity of the considered mechanisms varies between 0.09 g and 0.53 g. The lowest capacity corresponds to the partial mechanism of the facade with vertical cracks in the connections with the towers (Mechanism 5). This is an extremely conservative case and implies a very poor connection between the longitudinal and transversal walls. There is no current damage or reasonable suspicion that indicates a poor-quality connection between these walls.

Mechanism 2 (collapse of the belfry) results in a maximum capacity equal to 0.48 g, which is significantly higher than the capacity obtained from the pushover analysis in the positive longitudinal direction (0.29 g). However, it must be noted that the collapse obtained from the pushover analysis involves diagonal cracks in the columns of the belfry (Figure 12), which differs from the mechanism proposed for the limit analysis (horizontal cracks at the base of the belfry).

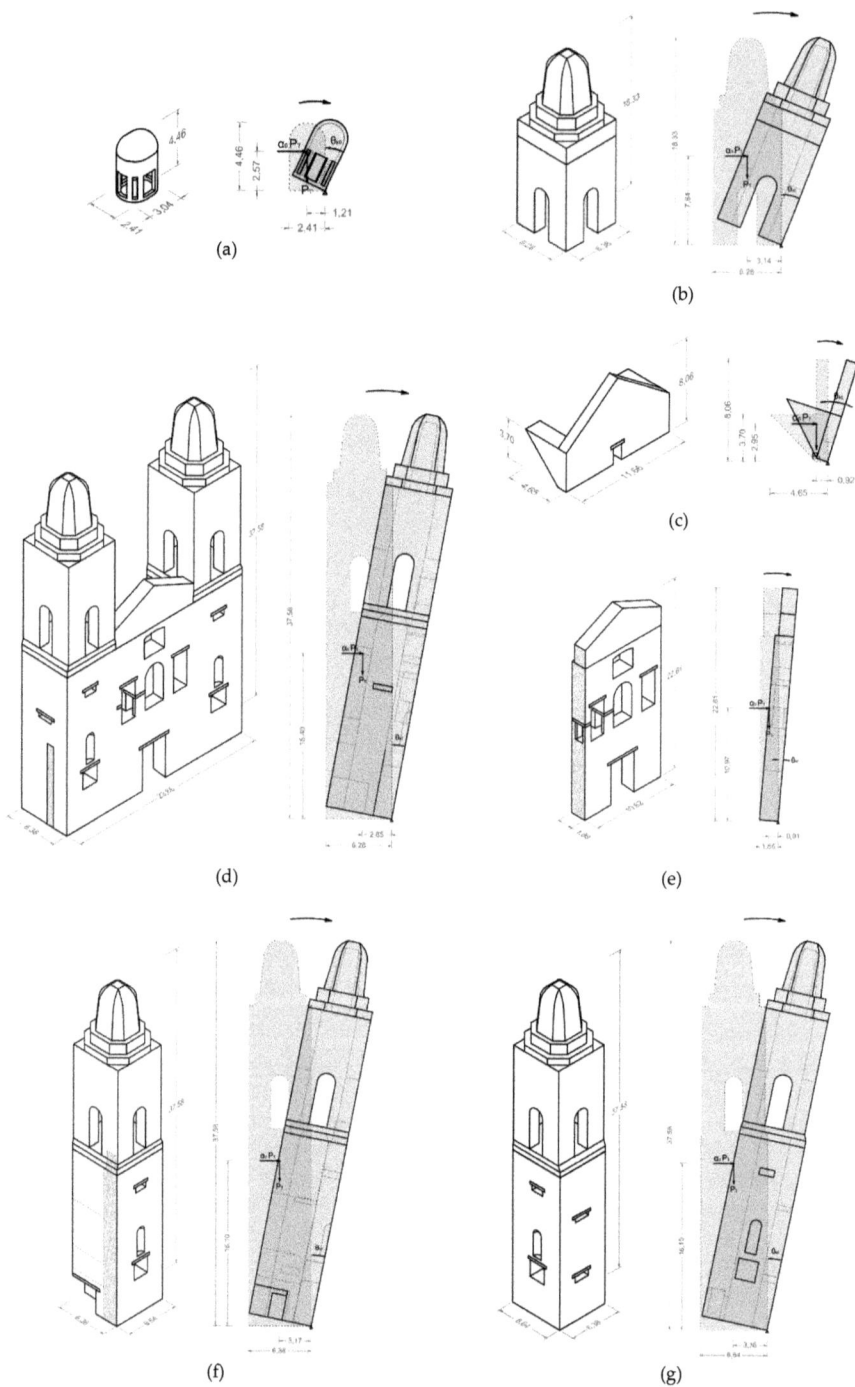

Figure 14. Collapse mechanisms considered for limit analysis: (**a**) Mechanism 1, (**b**) Mechanism 2, (**c**) Mechanism 3, (**d**) Mechanism 4, (**e**) Mechanism 5, (**f**) Mechanism 6, (**g**) Mechanism 7. Units in m.

Table 3. Safety verification for seismic action by means of limit analysis.

Mechanism	Maximum Capacity (g)	Required Capacity (g)	
		Type 1	Type 2
1	0.53	0.08 [B]	0.06 [B]
2	0.48	0.04 [B]	0.03 [B]
3	0.31	0.05 [B]	0.04 [B]
4	0.24	0.03 [A]	0.04 [A]
5	0.09	0.03 [A]	0.04 [A]
6	0.26	0.03 [A]	0.04 [A]
7	0.25	0.03 [A]	0.04 [A]

[A] Verification for mechanisms in contact with the soil. [B] Verification for mechanisms above the ground.

Mechanism 4 (overturning of facade and towers) corresponds to a failure mode also observed in the pushover analysis (Figure 13). In this case, the maximum capacity obtained from both approaches is the same (0.24 g). Additionally, Mechanism 6 (overturning of the tower in the direction parallel to the facade) results in a maximum capacity (0.25 g) similar to the one obtained by means of the nonlinear static analysis in the positive transversal direction (0.28 g). The lower value of the limit analysis could be explained by the differences between the failure mechanisms defined for one and the other methods.

Finally, the limit analysis proves that the collapse mechanisms considered in this study comply with the stability criteria for seismic actions defined in the Italian code NTC-08 (Table 3), with a minimum safety factor of 2.30. Overall, the limit analysis results in capacity values equivalent to those obtained by means of the nonlinear static analysis, despite some differences in the damage patterns associated with the mechanisms.

6. Conclusions

A 3D FEM model was prepared and validated in order to evaluate the performance and structural stability of the church of São Miguel de Refojos (Portugal). The seismic behavior was assessed by means of nonlinear static analysis as well as limit analysis. Moreover, the safety condition of the structure was verified according to the prescribed criteria established in national and international structural codes.

In the first stage, the inspection and diagnosis campaign allowed us to identify and characterize the main structural aspects necessary to evaluate the state of conservation of the building and generate the numerical model. From these works, it was concluded that the current structural condition and the overall state of conservation are good. Then, the model was calibrated with respect to the dynamic properties estimated from the dynamic identification tests. Subsequently, the numerical and kinematic analyses allowed us to conclude that the church meets the safety requirements prescribed for the seismic action. The results showed that the negative longitudinal direction (−YY) is the most vulnerable one and is associated with the overturning mechanism of the main facade and towers with rotation at the base (out-of-plane collapse). Nonetheless, this mechanism has a safety factor equal to 3.00.

Taking into account the assumptions and results of this study, the safety condition of the structure is guaranteed. However, a set of preventive measures should be put into practice in order to guarantee a more efficient structural behavior and the necessary works for the conservation of the building, namely the filling of existing cracks, repair of deformed floors, and development and application of a monitoring plan including the revealing aspects for the structural condition, such as regular visual inspection.

Author Contributions: Investigation, R.R., N.M. and P.B.L.; Validation, N.M. and P.B.L.; Writing – original draft, R.R. and P.B.L.; Writing – review & editing, R.R. & N.M.

Funding: This research received no external funding.

Acknowledgments: The authors would like to acknowledge the contribution of the Municipality of Cabeceiras de Basto and the Regional Northern Culture Directorate of Portugal for the data that they provided and their support.

Conflicts of Interest: The authors declare no conflict of interest.

References

1. Lemos, C.; Queiroga, F.; Vitorino, A.; Melo, L. *Remodeling of the Cloister of the Monastery of S. Miguel de Refojos, Cabeceiras de Basto*; Final Report of Archaeological Works; Perennia Monumenta: Oporto, Portugal, 2013. (In Portuguese)
2. Sequeira, M. The Church of the Monastery of São Miguel de Refojos in Cabeceiras de Basto. In *Studies in honor of Professor José Amadeu Coelho Dias*; Faculdade de Letras da Universidade do Porto (FLUP): Porto, Portugal, 2006; Volume 2, pp. 223–232. (In Portuguese)
3. Pereira, J.F. *Baroque Architecture in Portugal*; Portuguese Culture and Language Institute (Ministry of Education): Lisbon, Portugal, 1992. (In Portuguese)
4. PAUTA. *Study of the Edification of the Church and Sacristy of the Convent of Refojos*; Municipality of Cabeceiras de Basto: Cabeceiras de Basto, Portugal, 2000. (In Portuguese)
5. DGPC. *Church and Sacristy of the Monastery of Refojos, as well as the Ceiling of One of the Rooms of the Old Monastery of Benedictine Friars*; General Directorate of Cultural Heritage: Lisbon, Portugal, 2015. (In Portuguese)
6. Castellazzi, G.; Gentilini, C.; Nobile, L. Seismic vulnerability assessment of a historical church: Limit analysis and nonlinear Finite Element Analysis. *Adv. Civ. Eng.* **2013**, *1*, 1–12. [CrossRef]
7. Milani, G. Lesson learned after the Emilia-Romagna, Italy, 20–29 May 2012 earthquakes: A limit analysis insight on three masonry churches. *Eng. Fail. Anal.* **2013**, *34*, 761–778. [CrossRef]
8. Lourenço, P.B.; Mendes, N.; Ramos, L.F.; Oliveira, D.V. Analysis of masonry structures without box behavior. *Int. J. Archit. Herit.* **2011**, *5*, 369–382. [CrossRef]
9. Angelillo, M.; Lourenço, P.B.; Milani, G. Masonry Behaviour and Modelling. In *Mechanics of Masonry Structures*; Angelillo, M., Ed.; Springer: Milan, Italy, 2014; Volume 551, pp. 1–26.
10. ARTeMIS. *Modal Users' Manual*; SVS–Structural Vibration Solutions: Aalborg East, Denmark, 2014.
11. Ewins, D. *Modal Testing: Theory, Practice and Application*, 2nd ed.; Research Studies Press LTD: Baldock-Hertfordshire, UK, 2000.
12. Fernandes, F. Evaluation of Two Novel NDT Techniques: Microdrilling of Clay Bricks and Ground Penetrating Radar in Masonry. Ph.D. Thesis, University of Minho, Braga, Portugal, 2006.
13. Johnston, B.; Ruffell, A.; McKinley, J.; Warke, P. Detecting voids within a historical building façade: A comparative study of three high frequency GPR antenna. *J. Cult. Herit.* **2018**, *32*, 117–123. [CrossRef]
14. DIANA FEA BV. *DIsplacement Method ANAlyser, Release 10*; DIANA FEA BV: Delft, The Netherlands, 2017.
15. Gaetani, A. Seismic Performance of Masonry cross Vaults: Learning from Historical Developments and Experimental Testing. Ph.D. Thesis, University of Minho, Braga, Portugal, 2016.
16. NTC-08 Guidelines. *Technical Standards for Constructions*; Official Journal of the Italian Republic: Rome, Italy, 2008. (In Italian)
17. IMIT. *Circ. 02.02.2009, n. 617: Instructions for the Application of the New Technical Standards for Construction Referred to in the Ministerial Decree of 14 January 2008*; Italian Ministry of Infrastructures and Transportation: Rome, Italy, 2009. (In Italian)
18. Vasconcelos, G. Experimental Investigations on the Mechanics of Stone Masonry: Characterization of Granites and Behavior of Ancient Masonry Shear Walls. Ph.D. Thesis, University of Minho, Braga, Portugal, 2005.
19. Poletti, E. Characterization of the Seismic Behaviour of Traditional Timber Frame Walls. Ph.D. Thesis, University of Minho, Braga, Portugal, 2013.
20. Douglas, B.M.; Reid, W.H. Dynamic tests and system identification of bridges. *J. Struct. Div.* **1982**, *108*, 2295–2312.
21. Tomazevic, M. *Earthquake-Resistant and Design of Masonry Buildings*; Imperial College Press: London, UK, 1999.
22. CEB-FIB. Material Properties. In *CEB-FIP MODEL CODE 1990*; T. Telford: London, UK, 1993; pp. 33–81.
23. Lourenço, P.B. Recent advances in masonry modelling: Micromodelling and Homogenisation. In *Multiscale Modeling in Solid Mechanics: Computational Approaches*; Galvanetto, U., Ferri Aliabadi, M.H., Eds.; Imperial College Press: London, UK, 2009; Volume 3, pp. 251–294.

24. Mendes, N. Seismic Assessment of Ancient Masonry Buildings: Shaking Table Tests and Numerical Analysis. Ph.D. Thesis, University of Minho, Braga, Portugal, 2012.
25. Peña, F.; Lourenço, P.B.; Mendes, N.; Oliveira, D.V. Numerical models for the seismic assessment of an old masonry tower. *Eng. Struct.* **2010**, *32*, 1466–1478. [CrossRef]
26. O'Hearne, N.; Mendes, N.; Lourenço, P.B. Seismic analysis of the San Sebastian Basilica (Philippines). In Proceedings of the 40th IABSE Symposium: Tomorrow's Megastructures, Nantes, France, 19–21 September 2018.
27. EN 1998-1. *Eurocode 8: Design of Structures for Earthquake Resistance—General Rules, Seismic Actions and Rules for Building*; European Commitee for Standardization: Brussels, Belgium, 2004.
28. NIKER. *Inventory of Earthquake-Induced Failure Mechanisms Related to Construction Types, Structural Elements, and Materials*; POLIMI, D 3.1; Politecnico di Milano: Milan, Italy, 2010.

MDPI

St. Alban-Anlage 66

4052 Basel

Switzerland

Tel. +41 61 683 77 34

Fax +41 61 302 89 18

www.mdpi.com

Buildings Editorial Office

E-mail: buildings@mdpi.com

www.mdpi.com/journal/buildings

www.ingramcontent.com/pod-product-compliance
Lightning Source LLC
Chambersburg PA
CBHW051857210326
41597CB00033B/5929